EFFECTIVE
SECURITY
MANAGEMENT

EFFECTIVE
SECURITY
MANAGEMENT

Charles A. Sennewald, C.P.P.

Security World Publishing Co., Inc.
2639 So. La Cienega Blvd. / Los Angeles, CA 90034

First Edition 1978

Security World Publishing Co., Inc.
2639 South La Cienega Boulevard
Los Angeles, California 90034

Printed in the United States of America

Library of Congress Cataloging in Publication Data

Sennewald, Charles A   1931–
   Effective security management.

   Includes index.
   1. Industry—Security measures.   2. Retail trade—Security measures.
3. Management.   I. Title.
HV8290.S46   658.4'7   78–6058
ISBN 0–913708–30–5

To my family,
who, close behind faith,
are my most precious possession.

Contents

The Security Director As A Leader • As A Company Executive •
As An Executive With High Visibility • As An Executive With a
Broad Profile • As An Innovator • As A Counselor and Adviser •
As A Trainer • As A Contemporary Professional • As A Goal
Setter • The New Security Director

Ultimate Goal of Supervision: Performance • The Supervisor As
An Inspector • The Supervisor and the Individual Employee •
Supervisory Authority • The "In-Between" Man • Span of
Control • One Boss • Automatic Shifting in the Line of Com-
mand • Functional or Staff Supervision • Supervisory Training

Standards of Conduct in Security Industry • Courtesy • Responsibility • Due Process: Respect for the Rights of Others • Cooperation • Personal Integrity • Self-Respect • Honesty • Cleanliness • Stability • Fidelity • Morality • Attitude

PART II   SECURITY PERSONNEL MANAGEMENT

Recruiting Activity • Entry or First Level Positions • Non-Entry
Level Recruiting • Interviewing Preliminaries • The Interview •
Secondary Interviewing • Selection of the Best Candidate •
Background Investigation of Applicant • Job Offer

Shortcomings of Typical "Training" • Training Defined •
"POP" Formula: Policy, Objective, Procedure • Detailed Ex-
pansion of Procedures • Training as Ongoing Responsibility •
Types of Security Training Programs • Meeting Organizational
Needs • Security Training Manual

of Career Personnel • Disadvantages of Career Personnel • Advantages of Non-Career Personnel (In-House Part-Time) • Disadvantages of Non-Career Personnel (In-House Part-Time) • Advantages of Non-Career Personnel (Contractual) • Disadvantages of Non-Career Personnel (Contractual) • Combining Career and Non-Career Personnel

What Is A Budget? • Why Do We Have A Budget? • When Is A Budget Prepared? • Who Participates in the Budgeting Process? • Top-Down, Bottom-Up Process • How Is The Budget Prepared? • Salary Expense Budget • Sundry Expense Budgets • Justifying the Security Budget

The Inspection Process • Support of Senior Management • Continuous Inspections • Formal or Informal Inspections • Structured or Unstructured Inspections • Who Conducts the Inspection? • Assessment of Risks and Countermeasures • Risk Assessment • Selection of Countermeasures • Procedural Controls • Hardware • Electronics • Personnel • Assessment of Countermeasures • Inspecting for Compliance with Procedures • Statistics in Program Management

Investigative Skills Used in Secondary Assignments • Applicant Background Investigations • Internal Criminal Attacks • External Criminal Attacks • Corporate Integrity Investigations • Corporate Liability Investigations • Labor Matters • Miscellaneous Investigations • Establishing the Investigative Function • Undercover Investigation • Confidentiality of Investigations • Interesting Cases

Six Basic Functions • Description of Functions • Supervision •

The Public's Need for Information and the Company's Need to Inform the Public • Public Speaking • Suggested Topics for Various Audiences • News Press Appearances or Interviews • Radio Interviews • Participation in Community-Oriented Projects • General Public Contact

Part I
GENERAL
SECURITY
MANAGEMENT

Chapter 1

General Principles of Organization

The framework of structured organization is nothing more nor less than a vehicle for the realization of the purposes for which a company or a department is established. That skeleton, the organizational structure itself, does not think, has no initiative, cannot act or react. Yet it is absolutely essential in the work environment. A sound organizational framework facilitates the accomplishment of tasks by members of the organization, people working under the supervision of responsible managers.

A hospital, for example, is organized for the purpose of providing health care services. A sub-unit of that master organization, the Security Department, is organized for the purpose of protecting that health care environment. Organization, then, is the arrangement of people with a common objective or purpose, in a manner to make possible the performance of related tasks grouped for the purpose of assignment, and the establishment of areas of responsibility with clearly defined channels of communication and authority.

ORGANIZATIONAL PRINCIPLES

In the design of a sound organizational framework there are six widely accepted principles:
1) The work should be divided according to some logical plan.

2) Lines of authority and responsibility should be made as clear and direct as possible.
3) One supervisor can effectively control only a limited number of people, and that limit should not be exceeded. (This principle is called Span of Control.)
4) There should be "unity of command" in the organization.
5) Responsibility cannot be given without delegating commensurate authority, and there must be accountability for that authority.
6) All efforts of sub-units and personnel must be coordinated into the harmonious achievement of the organization's objectives.

Each of these principles has a meaningful application within a security organization.

Logical Division of Work

The necessity for the division of work becomes apparent as soon as you have more than one man on the job. *How* the work is divided can have a significant impact on the results at the end of the day. The manner and extent of the division of work influence the product or performance qualitatively as well as quantitatively. The logical division of work, therefore, deserves close attention.

There are five primary ways in which work can be divided:
- purpose
- process or method
- clientele
- time
- geography

Purpose. It is most common for work to be divided according to purpose. In fact, the Security Department could be organized into two divisions: a Loss Prevention Division (its purpose being to prevent losses) and a Detection Division (its purpose being to apprehend those who defeated the efforts of the prevention unit).

Process or method. A process unit is organized according to the method of work, all similar processes being in the same unit. An example in security might be the polygraph examiners unit or the credit card investigators unit of the general investigative section.

Clientele. Work may also be divided according to the clientele served or worked with. An example here would be the background screening personnel, who deal only with prospective and new employees; or store detectives, who concentrate on shoplifters; or general retail investigators, who become involved with dishonest employees, shoplifters, and a host of other criminal offenders.

Division of work by purpose, process or clientele is really a division based on the *nature* of the work itself and consequently is referred to as "functional." In other words, the grouping of security personnel to perform work divided by its nature (purpose, process or clientele) is called *functional organization.*

For a great many organizations, the functional organization constitutes the full division of work. Security, however, like police and fire services in the public sector, usually has around-the-clock protective responsibilities. In addition, unlike its cousins in the public sector, it may have protective responsibilities spread over a wide geographic area.

Time. At first glance the 24-hour coverage of a given facility may appear relatively simple. The natural reaction is to have three eight-hour shifts, with fixed posts, patrol, and the communication and alarm center all changing at midnight, 8:00 a.m. and 4:00 p.m. In reality a number of interesting problems surface when a department begins organizing by time, such as:

- How many security people are necessary on the first shift? If a minimum security staff takes over at midnight and the facility commences its business day at 7:00 a.m., can you operate for one hour with the minimum staff, or must you increase coverage prior to 7:00 a.m. and overlap shifts? (There are hundreds of variables to just this type of problem.)
- If you have two or more functional units, such as men assigned to patrol and men assigned to the communications and alarm center (in another organizational pyramid altogether), who is in command at 3:00 a.m.? The question of staff supervision confuses many people. (See Chapter 5, The Security Supervisor's Role, for detailed discussion of staff supervision.)
- How much supervision is necessary during facility down-time? If the question is not *how much,* then how is *any* supervision exercised at 3:00 a.m.?

- If there are five posts, each critical and necessary, and five men are scheduled and one fails to show, how do you handle the situation? Should you schedule six men for just that contingency?

These and other problems do arise and are resolved regularly in facilities of every kind. Organizing by time, a way of life for security operations, does create special problems that deserve consideration if organizing by time is a new undertaking for a company.

Geography. Whenever a security department is obliged to serve a location removed from the headquarters facility, and one or more security personnel are assigned to the outlying location, there is one major issue that must be resolved: to whom do the security personnel report—to security management back at headquarters, or to site management (which is non-security)?

The real issue is: Should non-security management have direct supervision over a security employee who has technical or semi-technical skills that more often than not are beyond the competence of that non-security management?

In defining the type of authority an executive or supervisor exercises, a distinction is generally made between *line* and *staff* authority. While these terms have many meanings, in its primary sense line authority implies a direct (or single line) relationship between a supervisor and his subordinate; the staff function is service or advisory in nature.

Security personnel should only be directly supervised by security management. Site management may provide staff supervision, providing suggestions and assistance, but restricted to such matters as attention to duty, promptness in reporting, and compliance with general rules. Detailed security activities fall outside the jurisdiction of such a manager.

Non-security management should not have line authority (direct supervision) over security, not only because of the issue of professional competency, but also because site management should not be beyond the "reach" of security. Site management is indeed "out of reach" if the only internal control, security, is subject to its command. Site management would be free to engage in any form of mischief, malpractice or dishonesty without fear of security's reporting the activities to company headquarters.

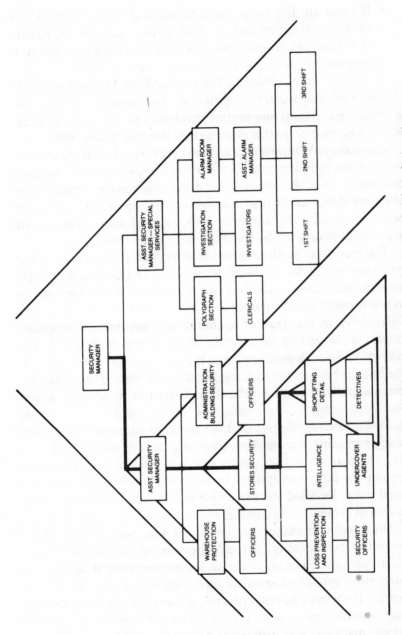

Figure 1-1. Repeated pyramidal forms within organizational structure.

Clear Lines of Authority and Responsibility

Once the work has been properly divided, the organization takes on the appearance of a pyramid-like structure, within which are small pyramids, as illustrated in Figure 1-1. Each part of each pyramid defines, with exactness, a function or responsibility and to whom that function is responsible. One can easily trace the solid line upwards to the Security Manager or Director who is ultimately responsible for every function within the security organization.

Not only is it important to have this organizational pyramid documented, normally in the form of an organizational chart, but it is also essential that security employees have access to that chart so they can see exactly where they fit into the organization pattern, to whom they are responsible, to whom their supervisor is responsible, and so on right up to the top. Failure to so inform employees causes unnecessary confusion, and confusion is a major contributor to ineffective job performance.

Additionally, the organizational chart is a subtle motivator. People can see themselves moving up in the boxes; in order for goal-setting to be successful, one must be able to envision oneself already in possession of one's goal.

Finally, the apparent rigidity of boxes and lines in the organizational chart must not freeze communications. Employees at the lowest layer of the pyramid must feel free to communicate directly with the Security Manager, and have the right to do so without obtaining permission from all the intervening levels of supervision.

Span of Control

There is a limit to the number of subordinates who can be supervised effectively by one person, and that limit should not be exceeded. The limit ranges from a maximum of five at the highest level in the organization, to a maximum of twelve at the lowest level. The greater the degree of sophistication of the interactions between supervisors and subordinates, the narrower is the optimum span of control.

Exceeding the limits of span of control is really no different from spreading oneself too thin in some non-work environment such as school. If a student carries a full academic load of core subjects, becomes involved in student government, goes out for varsity football, is engaged to be married, belongs to the military reserve, and works twenty hours a week in a liquor store, it is likely that some of his activities will be slighted and very few of them, if any, will be done with excellence.

Slipshod, undisciplined, poorly executed security work is an almost inevitable consequence of violating the organizational principle of span of control.

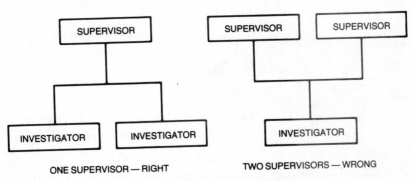

ONE SUPERVISOR — RIGHT TWO SUPERVISORS — WRONG

Figure 1-2. Unity of Command.

Unity of Command

The fourth principle, that of unity of command, means that an employee should be under the direct control of *one and only one* immediate superior. This principle also dictates that a task or function requiring the action of two or more people must also be under the direct control of but one supervisor.

Violations of the principle of unity of command are not usually found in the design of the organization, but rather occur, more by accident than design, during special events or other non-routine conditions that bring out more company executives than usual. The security officer or agent who is given conflicting orders by several superiors becomes confused, inefficient, irresponsible and, to say the least, angry.

One cannot successfully serve two masters. This will be discussed further in Chapter 5.

Responsibility, Authority and Accountability

The fifth principle of organization is all too frequently violated by the manager or executive who gives a subordinate the responsibility to do a task but holds back the authority needed to discharge such responsibility effectively. A prime example of this is the case where a man is given the job of supervising a unit. He is told the unit is his and he will be measured by how well he performs in that assignment. The new supervisor then discovers he does not have the right (authority) to select the applicant of his choice to fill a vacancy; he discovers that disciplinary matters are decided by his superiors (and his subordinates know this); and he soon finds that his plans, suggestions and ideas are replaced with those of his boss, and thus he is totally frustrated. He has a job and yet he does not.

The true art of delegation includes giving responsibility with commensurate authority and then holding the employee fully accountable for his use of that authority. *If,* in the above example, the new supervisor had the authority to hire, discipline, and implement his own ideas, and did so without exercising good judgment, and could not be corrected or trained to use his authority properly, then he should be stripped not only of the authority but of the responsibility as well. The employee must be given *both* responsibility and authority and held accountable for both.

Perhaps the major reason why so many managers violate this principle is that they are unwilling, sometimes subconsciously, to allow subordinates to carry out the responsibilities delegated to them, because the manager knows that he is ultimately responsible. The manager knows it is true that you cannot completely delegate responsibility. This may seem confusing and sound like double-talk, so let us approach the problem from a different angle.

   a) The head of the security pyramid, the Security Manager or Security Director, is the only one accountable for his organization, the Security Department. His reputation grows brighter in the department's successes and suffers in its failures. Almost invariably, he rose to the top because of his proven ability and track record. In other words, doing things his way has proven, over the stretch of time, to be successful. That is why he is where he is.

b) The security manager or director cannot do the entire security job alone and needs people to help him get it done. Depending on the scope of the job, he may need anywhere from two people to three hundred people. Ideally, if every person on his team thought and acted exactly as he does, he would assure himself of continued outstanding success.

c) The manager understands, however, that no one else thinks and acts exactly as he does. He may reason that the next best thing, then, is to do the thinking for all the key people, make the decisions for them, have them run the organization his way. He holds the authority, and when things go wrong— and they will—he will probably severely criticize the party who failed. This manager has not really delegated responsibility and authority. Yet, ironically, when things go wrong he will point out to management the employee who failed, and in so doing will have someone to *share* the responsibility with, because ultimately it is his responsibility.

d) Or, in contrast to the situation described in (c) above, the manager may open up the organization to other talented people. Within the broad guidelines he sets as a leader, he will give those key people genuine responsibility. He will make them accountable, and they will respond positively to that accountability. When things go wrong—and they will— the party who failed will judge himself critically. This manager *has* truly delegated responsibility and authority. Yet, ironically, he will take full responsibility for failure, because it *is* ultimately and rightfully his responsibility.

Coordination to Meet Organizational Goals

Theoretically, if the first five principles just discussed were adhered to, everything should function smoothly. In practice such total harmony is rare, if not impossible. Human frailties such as jealousy over assignments as well as promotions, elitism in some sub-units, friction between supervisors, the historic poor reputation of certain sub-units or assignments—all these and more tend to decrease what might have been optimum efficiency.

How then does management coordinate all efforts of sub-units and personnel? Or, better, what can management do to *attempt* to coordinate all units and personnel into the harmonious achievement of the department's goals? The answer is: establish a sound security training program and good departmental communications.

Both training and communications are dealt with in separate chapters (Chapters 8 and 12, respectively). The emphasis in both cases should be on educating employees about the organization and its objectives; defining the importance of each sub-unit's contribution to the whole; developing organizational pride and individual security employee self-esteem; creating a sense of security unity and identification within the company as a whole; and, finally, developing a climate wherein the individual employee includes organizational goals within his own personal goals.

WHERE SECURITY FITS
IN THE ORGANIZATIONAL STRUCTURE

The Changing Role of Security

In less than two decades the security function has climbed from the depths of organizational existence, from dank and smelly basement offices, to the heights of executive offices and a place in the sun. Gone for most organizations are the days when the Security Department reported to the superintendent of buildings, manager of buildings and grounds, head housekeeper, yardmaster, controller, plant engineer or store manager. Today security management frequently reports directly to senior management, if not the chief executive officer himself. And that vertical movement is still gaining momentum. Security is moving more and more into the ranks of senior management; more positions such as Vice President of Security, Vice President of Loss Prevention or Vice President of Assets Protection are being created yearly.

Why this sudden ascent? Simply stated, the ever-increasing contribution security makes to the organization's objectives, principally profit, has earned corresponding increased recognition from top management.

Another reason for the growing recognition of security's importance is the increasing prevalence of crime in our society. A number of socio-economic factors, along with political and cultural conditions, have combined to create a climate in which deviant or anti-social behavior is no longer intolerable. As a result, more and more deviant behavior occurs, particularly attacks against property (theft) —until the magnitude of the problem far transcends the limited prevention ability of public law enforcement. The burden of crime prevention falls on the private sector.

In an address to the International Association of Chiefs of Police, Richard W. Velde, Administrator of the Law Enforcement Assistance Administration, stated:

> The criminal justice system, and particularly our nation's police, do perform a rather narrow function that is largely a responsive one that follows the commission of crime. There are constitutional and statutory responsibilities in all the states that define the role of the police force and essentially they say that police are not in the crime prevention business.*

Without question, a large number of firms and even entire industries would fail today without their own internal security organizations. Imagine the position of a major credit card company without its security department. Who would coordinate and track the criminal abuse of that credit privilege across the country?

Security's Contribution to Profits

Security contributes to company or corporate profits by reducing or eliminating preventable losses, including those caused by criminal behavior. Consider the retail industry, for example. A major chain with sales of $1 billion might realize a 3 percent net profit as well as a 3 percent inventory shrinkage (these three figures are quite realistic). This firm, then, realizes $30 million in profits and $30 million in losses, or lost profit. If the Security Department through its efforts and programs can reduce the inventory shrinkage by just one-half of one percent, profits would rise $5 million!

*Velde, Richard W., quoted in *Private Security: Report of the Task Force on Private Security.* (Washington, D.C.: National Advisory Committee on Criminal Justice Standards and Goals, 1976), p. 19.

Where else can management find such opportunities to increase profits? The cost of raw materials cannot be reduced; they are becoming scarcer and more expensive. The cost of labor cannot be reduced; labor's demands are only going up. The costs of so-called fixed expenses such as rent, utilities and insurance cannot be reduced; they are all rising. Because losses are so enormous, their reduction is in the hands of protection professionals who manage corporate and divisional security organizations.

To Whom Does Security Report?

With increased recognition of the need for security within the whole spectrum of company activities, all concerned directly or indirectly with the "bottom line" of business (profits), came increased responsibility; and with increased responsibility came commensurate authority. To provide the security manager with that necessary authority, he along with his organization has moved up in the organizational pyramid to report directly to senior management, usually a vice president. That vice president delegates a portion of his authority to the security chief, who can then exercise what is known as *functional* authority.

Reporting directly to a vice president places the security executive at the top of the middle management sector of business. The most dynamic people in the firm are now his peers. Figures 1-3 and 1-4 illustrate the place typically held in the organizational chart by a Security Director in a manufacturing and a retail organization, respectively.

The Difference Between Corporate and Company Security

Although the word "corporate" is sometimes used to describe a firm's central authority, the word more accurately refers to that small holding organization that owns a number of firms. A conglomerate is a combination of a variety of individual companies, each with its own executive team, its own goals, its own volume and its own profit performance. The financial results of each of the companies in a corporate structure (or conglomerate) are, for the sake of simplicity, forwarded to the corporate organization at the very top of the pyramid.

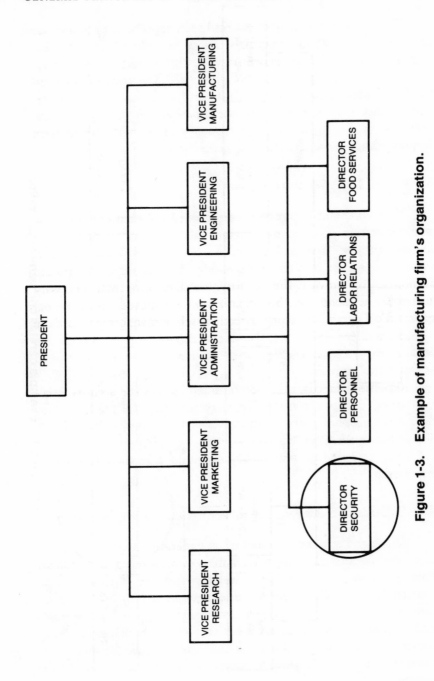

Figure 1-3.   Example of manufacturing firm's organization.

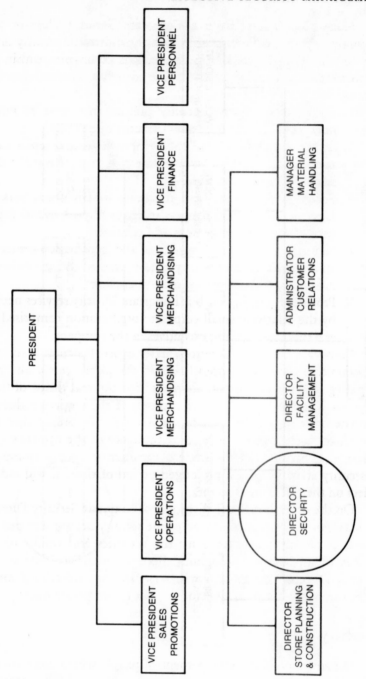

Figure 1-4.   Example of retail company organization.

Many corporations have a Corporate Security Director who maintains a pure staff relationship with the individual security directors in charge of protection in the various companies within the corporate family. (See Figure 1–5.) The corporate director's job is as follows:

1. Establishes corporate security policies that serve as guidelines for divisional (company) security operations.
2. Serves as an advisor and counselor to divisional senior management in terms of his assessment of how effectively divisional security is functioning.
3. Serves as an advisor and counselor to Divisional Security Director, giving him support in terms of professional expertise, advice, encouragement and criticism.
4. Serves as a central clearing-house and information center for all divisions within the corporation, disseminating important information about the industry as a whole.
5. Provides for those few but important security services needed by the relatively small corporate organization comprised of, as a rule, top-ranking executives in the company.

Corporate Security Directors have other functions, such as maintaining liaison with top officials in the public sector and participating in trade association activities. But the real thrust of these corporate security jobs is one of counsel. If the corporate director has a security staff, it is usually quite small. He simply does not have direct accountability for the performance of the divisions; yet, certainly, if divisions demonstrate a consistently poor performance in security activities over a prolonged period of time, it will indeed reflect on the corporate director.

On the other hand, the divisional or company Security Director (or Manager or whatever the head of the security department may be called) is directly accountable for the activities and results of the security organization. Throughout this text, when we refer to the Security Director, unless otherwise specified, we are talking about the divisional or company director . . . not the corporate man.

Summary

Organization is the arrangement of people with a common objective in a manner that groups related tasks, establishes areas of

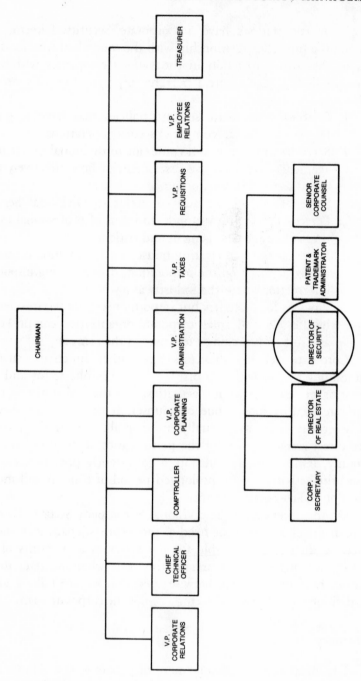

Figure 1-5. Example of corporate (conglomerate) organization.

responsibility, and defines lines of communication and authority.

There are six basic principles of organization: (1) logical division of work (according to purpose, process, clientele, time or geography); (2) clear lines of authority, visible on the organizational chart as a pyramidal structure; (3) limited span of control; (4) unity of command; (5) true delegation of responsibility and authority, with attendant accountability; and (6) coordination of efforts through training and communication.

Within the company or corporate organizational structure, security in recent years has shown a sharp vertical movement, an ascent primarily attributable to rising crime and increased recognition of security's contribution to profits. The Security Director now commonly reports to a member of senior management.

In the corporate or conglomerate structure, the Corporate Security Director serves generally in a staff relationship both to higher management and to the individual company Security Directors. In this text, discussion of the Security Director's or Manager's role refers to the security function in the individual company rather than that of the corporate organization.

Review Questions

1. Explain the five methods of dividing work.
2. Discuss the problems that may arise in organizing work by time.
3. Give two reasons why non-security management should not have line authority over security employees.
4. What is meant by *Span of Control?*
5. Explain the principle of *Unity of Command.*
6. Discuss the relationship between responsibility and authority. Give an example where a manager has given a subordinate responsibility without commensurate authority.
7. How does security contribute to the company's profits?

Chapter 2

Organizational Structure

The organizational structure of a department within a company will reflect the six general organizational principles discussed in Chapter 1, including:

- the logical division of tasks or responsibilities
- clear lines of authority and responsibility—within the department specifically and within the organization generally.

The department's organizational structure is two-dimensional in its formal representation, as illustrated in Figure 2–1. On the *horizontal* plane it indicates the division of areas of responsibilities; on the *vertical* plane it defines levels of authority or rank. In the illustration shown, responsibility for security under the Security Manager has been divided into two areas, with an Assistant Manager for Loss Prevention and an Assistant Manager for Investigations. The horizontal division defines areas of responsibility for each Assistant Manager, while the vertical chart indicates that they are of equal rank, each reporting directly to the Security Manager. Similarly, Loss Prevention responsibilities are divided among the officers in charge of the Computer Room Detail, General Facility Protection, and Credit Card and Accounts Receivable. The organizational chart indicates this separation of duties, as well as showing the relationship of each officer-in-charge of these subdivisions to the officers under them and the Assistant Manager and Security Manager above them.

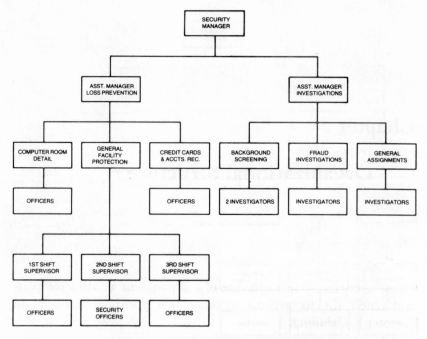

**Figure 2-1.   Formal organizational structure
of a security department.**

The organizational structure of one company's security department will differ in widely varying degrees from that of another company, *even within the same industry.* This is true for several reasons.
First, organizational structures are man-made, reflecting the way in
which the security director and management above him perceive
departmental and company priorities. The formal pattern will also
be influenced by individual personnel. Finally, the structure is and
must be fluid, or highly changeable, to meet the ever-changing
character of most private enterprise operations.

The Informal Organization

It is revealing to compare the formal organizational structure
outlined in Figure 2–1 with the informal organization in Figure 2–2,
which illustrates how the typical organization might really work.
The conspicuous difference between the two organizations is

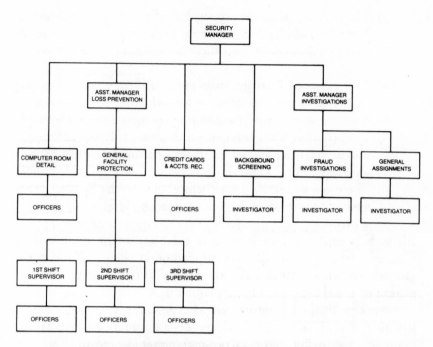

**Figure 2-2.   Sample "informal" organizational structure
of a security department.**

that, in the informal organization—the *real* organization—five supervisors report directly to the Security Manager, not just two. In addition to the two Assistant Managers, there are the supervisors in charge
of the Computer Room, Credit Card and Accounts Receivable, and
Background Screening. Reasons behind such a change in structure
will be many. They might include any of the following:

1) Physical protection of the computer room and of the credit
   card and accounts receivable areas logically belong under
   the Loss Prevention banner, but that Assistant Manager has
   limited interests and talent, favors the area of general facilities protection, and as a consequence has literally abrogated
   the other two functions.

2) The supervisors of those two areas of responsibility do not
   accept the Assistant Manager for Loss Prevention as their
   immediate superior and look instead to the Security Manager, who permits this condition.

3) The Security Manager has little confidence in that Assistant Manager's ability but is unable to fire him, with the result that he informally brings those sensitive units under his own wing.
4) The Security Manager has great difficulty in delegating authority in sensitive areas to subordinates; consequently he tends to exercise direct authority over most activities himself.
5) The Security Manager regards such areas of activity as "toys," and he violates the organizational structure because he enjoys "playing" these games.
6) There is a subjective relationship between the Security Manager and those supervisors on that second level of supervision, a friendship that interferes with the organizational integrity.

All such possible explanations—and there are many more—could also, of course, explain why the supervisor in charge of the Background Screening Unit also reports directly to the Security Manager instead of to his logical immediate superior.

Another major factor in the changeability of organizational structure is company (and therefore departmental) growth. As an example, one major retail chain had eleven stores in 1961, all located in Southern California. Sixteen years later there were 45 major department stores and two clearance centers in five states, with other stores scheduled to open. Obviously, the structure of the security department was far removed from what it had been sixteen years ago. During that period the department experienced at least a dozen reorganizations. Contrast this flexibility with, for instance, a municipal police department with a relatively stable city population. Few, if any, major departmental reorganizations would be expected to occur during the same period of time.

Budgetary considerations also play an important role in the organizational design. As an example, consider again the organization illustrated in Figure 2-1. The chart indicates a total of six supervisors at the third level reporting directly to the two Assistant Managers. What would happen if the department's budget would allow only four supervisors at that level? How would the organization be changed?

A number of variations are possible. The Computer Room detail might be combined with the Credit Card and Accounts Receivable area into one unit or detail called the High Risk Detail, under one supervisor; similarly, the Background Screening unit might be combined with the General Assignments unit for investigations. Such horizontal shrinkage may or may not serve the best interests of the organization; however, budgetary restrictions may make such changes inevitable.

Whatever the changes required by growth or budget, the point of organization remains the same: to serve the interest of the department in getting its job done through an intelligent division of tasks and the establishment of clear lines of authority. This applies to the small organization as well as the larger one. There will be fewer vertical levels of authority or rank, and a simpler division of responsibilities on each horizontal plane, in a small department; but the *purpose* of organization and the approach to organizational structure are identical.

As we have already indicated, the structure is two-dimensional. In the ideal situation, achieving a viable organization involves three steps: (1) identify the departmental objectives, (2) identify the various tasks and divide them into logical work units, and (3) identify the levels of leadership necessary to achieve the tasks. All that remains is to "fill in" the boxes with appropriate personnel.

This is the ideal. Unfortunately, it does not normally work that way. As a rule, people—the employees to be put into the boxes— are already aboard. Thus the two-dimensional plan of organizational design becomes complicated by the introduction of what might be called a third dimension—that being, of course, the personnel. The results are almost inevitably bad because the design tends to lean towards personnel considerations. Expressed in another way, there is a tendency to build jobs, and organizations, around people, rather than identifying qualified talent and placing them in the jobs defined by a plan of organization. This is true because, for policy or personnel reasons, the security manager in charge of a department for the most part has no choice but to make the best use he can of existing, in-house personnel. It is easier to change the organization than it is to change the man.

The reality of organizational structure, then, is inevitably a compromise between a pure design, based on the best possible horizontal and vertical layout, and existing security department employees. For this reason the typical organizational chart must be suspect.

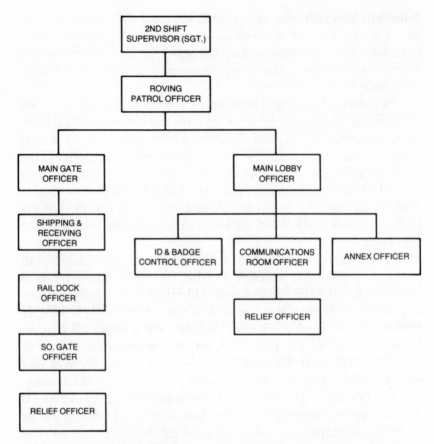

**Figure 2-3. Sub-unit's structure within security department's
organizational structure. Such mini-structures
may exist without official departmental recognition
or planning.**

The organizational chart is suspect in another significant way: it
clearly defines reporting lines, or "chain of command," when, in
reality, numerous informal reporting lines may exist. This aspect
of organizational life clouds the levels of authority, taking from
some and adding to others. In this respect the three-dimensional
aspect of the organization acquires an almost sinister air in terms of
who really is whose boss and who really can tell whom what to do—
and who is meddling in areas outside his scope of responsibility.

Sub-Unit Structures

In addition to the departmental organizational structure, both formal and informal, the security administrator should be aware of the existence of interdepartmental or sub-unit structures at lower levels within his own organization. The number and size of these mini-structures will be related to the size of the department. These structures will tend to have the same characteristics as the formal organization—that is, horizontal divisions (of tasks) and vertical levels (of authority).

Singling out the General Facility Protection division of our example department, consider the security officer's level on the second shift. To begin with, the second shift itself will probably represent the top of the nearly invisible hierarchical structure of security officers assigned to that division; the officers lower on the totem pole will be assigned to the least desirable shift, the first shift. The organizational structure of that second shift is shown in Figure 2–3.

In the absence of the Shift Supervisor, or Sergeant, the officer assigned to Patrol assumes leadership (in this example only). The Main Gate Officer assumes leadership for exterior protection, and the Main Lobby Officer is responsible for all security activities of an internal nature.

On the exterior protection side of this organizational pattern is a typical example of a "pecking order," or vertical line of authority; on the interior protection side, by contrast, all officers except the relief officer are of equal rank. In the latter case, in the absence of the Main Lobby Officer a struggle could ensue for position in the line of authority.

As a rule, such mini-structures of organization exist even though they have never been agreed upon by management, never been reduced to writing or charts, and in some cases never even been understood. Obviously, then, it is important for the security manager to be sensitive to the existence of such informal "structures" at the lowest levels of his organization, in order to maintain overall organizational integrity and harmony.

Consider this: How many times does a new supervisor, especially one who is not intimately familiar with the organization and its personnel, come onto the scene and commence changing operations and personnel around—and meet stubborn resistance? From an administrative point of view the consequences of putting a lieutenant over a captain are clearly understood. The same is true at the bottom of the organizational ladder; these employees have their own "captains" and "lieutenants," even if they were not officially appointed as such.

Sensitivity to such organizational facts of life not only avoids disharmony—a negative advantage—but, on the positive side, such insight can be highly productive in terms of organizational performance. If a superior officer or manager wants to insure that something is done in the absence of the Shift Supervisor, he goes to the Patrol Officer. If, instead, the manager bypasses the Patrol Officer and goes to the Main Lobby Officer, the job may not get done properly.

Summary

Organizational structure, then, as applied to a security department, is a valuable and necessary management tool to organize tasks and people in an intelligent, meaningful and responsible structure in order to meet, and successfully discharge, the security function in any company.

In the ideal this structure is two-dimensional, defining responsibilities (horizontal) and lines of authority (vertical). In practice the structure will be affected by a third dimension: personnel. This dimension is reflected in the influence of the individual security manager, the necessity of using existing personnel, and the presence of invisible interdepartmental structures.

The organizational structure, finally, is not and should not be rigid, since it must be capable of adapting to budgetary considerations, changing goals, and company size and growth.

Review Questions

1. What does the *horizontal* plane of an organizational chart represent? The *vertical* plane?
2. Why does the organizational structure of one security department differ from that of another, even within the same industry?

3. Discuss four reasons why the *informal* organization may differ from the *formal* organization.
4. What are the three steps in achieving a viable organizational structure?
5. Explain how personnel form the "third dimension" of organizational design.

Chapter 3

Security's Role in the Organization

THE PROTECTIVE SERVICE ROLE

The singularly most conspicuous role of the Security Department in any organization is that of Protector or Guardian—protecting the company's property, product or merchandise, assets, equipment, reputation, and employees. Such protection constitutes a service to the organization; thus the security department's function is one of service.

The value of such service is better measured by what does *not* happen than by what does. For the company to operate over a given period of time without a payroll holdup, major burglary, significant disappearance of inventory, equipment or documents, or the rape of a secretary in the parking lot is indicative of the security function's effectiveness in its guardian role. And the posture of the guardian role is one of prevention—prevention of crime and prevention of losses by means of a strategy and philosophy of denying the criminal the opportunity to succeed. In keeping with that posture, some organizations have abandoned the name "Security Department" in favor of "Department of Loss Prevention."

Stated another way, security could easily be identified as a *protective service of prevention*.

It is beyond the scope of this text to discuss in detail all of the individual protective duties with which a security department will become involved. The specific responsibilities of any security department will be adapted to the specific organization—its buildings and curtilage, its operations, its assets, its personnel, its interaction at all levels with the public, and its general environment. Nevertheless, the role of security normally involves common elements, among which the following could be included:

- Arrests and causes prosecution of all persons committing a criminal attack against company property, equipment, supplies, products, goods and/or other assets.
- Designs and implements physical controls of the facility.
- Administers and conducts identification badge program.
- Conducts pre-employment and post-employment screening.
- Monitors control of DOD classified documents and information.
- Maintains liaison with local, state and federal law enforcement authorities.
- Monitors controls of company proprietary information.
- Administers employee and visitor access to facility.
- Administers vehicular access and parking controls.
- Administers company's lock and key control program.
- Conducts security indoctrination and training.
- Investigates all criminal activity committed on company premises or against company interests, including attacks against persons.
- Administers executive protection program.
- Conducts financial stability investigations of potential vendors, merger candidates, etc.
- Coordinates special protection arrangements necessary during riots, natural disasters, strikes, etc.
- Designs and conducts security/loss prevention vulnerability surveys.
- Contracts for and administers outside security services such as guard services, undercover agents, shopping services, certain investigative services, polygraph services, armored transport services, document destruction, etc.
- Provides emergency courier and escort services as needed.
- Acts as consultant to senior management on all security-related matters.

Even this extensive list by no means exhausts the possible protection services that will fall upon a given security department, responding to particular hazards. What the above catalog does suggest is the general purpose of the security function in any organization: to protect the company (people and assets) against attack or loss.

Within security's protective role, there are a host of sub-roles which are often neglected or unrecognized by security management. These sub-roles may be divided into three service categories:

- Special Services
- Educational Services
- Management Services

SPECIAL SERVICES

The Security Department's objectives are designed to contribute to the achievement of company goals. Company executives, who provide vital leadership for company goal achievement, have personal goals which are difficult to separate from company goals. More often than not, their goals *are* company goals. Service, then, to the "company" and service to management should be synonymous, for what is good for the executive team is good for the company, and vice versa. All demands for protective service, whether clearly related to the work environment or of a peripheral nature involving senior management, require attention.

The security management that understands the reasonableness and logic of providing the broadest possible range of special services moves the security function more closely to the mainstream of the business and makes a more significant contribution to the overall success of the company. A sampling of special services follows.

Executive's Home Security Surveys

The executive who wishes to "harden" his home, installing protective measures against criminal intrusion and attack, has the choice of calling the police for advice, hiring an outside security consultant, attempting to select appropriate defenses himself, or calling on the company's security staff. The latter is recommended—providing, of course, that the staff has the expertise to achieve the desired degree of security.

The homes of executives are by far more attractive targets for burglary than those of the average employees, and it only makes good sense that extraordinary precautionary measures be taken.

In one recent case, a personnel executive's rented home in the hills of Southern California was burglarized several times in a like number of weeks. The question of increasing protection or asking for company security advice never crossed the executive's mind until total frustration set in. On the occasion of each burglary, the home was entered and ransacked while the executive was away, with no evidence of forcible entry. On each occasion the police came to the home and conducted their investigation, usually a surface examination of the physical premises and a documentation of pertinent facts surrounding the loss. The police and the executive theorized that a friend of the former tenants was still in possession of the house key and was responsible for the crimes. The former tenants had moved out of state, and the identity of the friend was unknown.

Based on that theory, the executive purchased a manual burglary alarm and attached it to the door. If entry was made through the door and the alarm would sound, the subsequent examination of the device would indicate it had been activated.

The home was again burglarized and the alarm was not activated. The executive was more than distraught over the dilemma. He finally turned to the company for assistance, and there, at his home, the security man within minutes located the point of entry used by the burglar—the louvered glass windows over the kitchen sink. The security executive temporarily secured these windows and the similar windows over the breakfast nook, and advised the executive of two alternative methods to secure those windows permanently. The executive was never victimized again. The simple technique of removing the screen and then removing the louvered window panes one at a time, and later replacing them, had escaped the police, literally as well as figuratively.

Executive home surveys will also examine the possible use of digital or central station alarming, inventorying valuable personal property (which includes recording serial numbers, photographing and/or marking); establishing emergency procedures; and exterior lighting, to name but a few of the areas of concern, depending upon the person and properties to be protected.

Investigative Assistance

Sooner or later the whole spectrum of investigative skills can be used in peripheral service—from tracing the license plate of a hit-and-run driver who sideswiped an executive's auto, to tracking the source of an obscene letter sent to an executive's home, to locating (in cooperation with police) the runaway daughter of an executive. And such investigative service need not be limited to executive or senior management problems; someone in middle management or a key supervisor in the company could have a problem that senior management felt was deserving of company attention.

Bodyguard/Escort Service

So-called bodyguard duties constitute another dimension to the variety of special services the security organization can provide. Such service could be any of the following:

1. Serving as an executive's chauffeur, temporarily or permanently.
2. Serving as escort for the dignitaries who are guests of the firm.
3. Serving as escorts for company executives visiting locations deemed hazardous.
4. Serving as escort for members of executives' families.
5. Serving as bartenders at special functions.
6. Intermingling with guests at special functions.
7. Escorting couriers or messengers.
8. Serving as courier.

Emergency Service

Most security departments run a twenty-four-hour-a-day operation from either an alarm room, security operations room, or desk. Because of that twenty-four-hour telephone capability, the department can offer company management a unique emergency service, as follows: Every member of management participating in the emergency plan provides the department with a data card that lists the name, sex and date of birth of the executive and his entire family;

their home address and phone number, with directions how to reach the home; the address and phone number of any summer or "second" homes, and directions how to reach those residences; the names and phone numbers of family physicians and dentists; local police department's address and phone number; local fire and rescue departments' names and numbers; local ambulance data; local hospital and emergency service data; insurance agent's identity and number; description and license numbers of family vehicles; identity of people to call (family, neighbors or friends) in the event of an emergency; etc.

That card is maintained in an alphabetical file in the twenty-four-hour operational room. When an executive is traveling, he calls in a supplement to the file, listing his itinerary, with phone numbers.

The emergency service becomes a clearing-house for processing emergency messages, dispatching emergency services, notifying appropriate people of problems, as well as expediting the flow of such information. Certainly the executive or a member of his family can call the police, fire, rescue squad, etc., directly and perhaps faster than routing the call through the Security Department. On the other hand, youngsters at home alone, or domestic servants, could be at a loss as to whom to call. Even an executive's wife might choose to call Security before calling the police if her husband is traveling— for example, if she became frightened and felt a prowler was on the property. A call from the company's Security Department reporting to the police that a prowler was on the property of an executive's home would, in all probability, receive quicker response than the semi-hysterical housewife calling the police herself.

There are a number of clear advantages to this type of service— still relatively unheard of and rare, but a growing function in the future of service-minded security departments throughout the country.

EDUCATIONAL SERVICES

An increasingly important and relatively new role for the Security Department is that of trainer and educator. As the private sector assumes more and more responsibility for law and order on private premises, there is an increasing need to educate employees and non-employees alike on the necessity and objectives of security.

A striking example of the need for employee security education is in the retail industry. Retailers, including food, drug, department, chain, discount, specialty stores and independents, lose millions of dollars each year to dishonest employees. Part of that loss is directly attributable to the fact that the employee is ignorant of the company's security efforts and capabilities to detect dishonesty.

Every day, new or relatively new employees ''discover'' clever methods to misroute or deliver merchandise or funds into their personal possession, unaware of the fact that the ingenious scheme has been attempted and detected thousands of times before. Because they have not been properly educated, they contrive for unwarranted advantages in total ignorance, damaging their employer and exposing themselves to the tragic consequences of detection, termination and prosecution—all for the want of a security induction or awareness program for new employees.

Who is to conduct such training sessions? Experience tells us that the most effective presentations on the security function are made by security personnel. They know what they are talking about, and their expertise is apparent. A security presentation by a training officer or member of management lacks the same degree of conviction or credibility. Security must, therefore, assume the role of trainer/educator.

New employee induction programs are but one of a number of educational activities Security is involved in.

General Security Programs

Whereas the induction training addresses itself to the new employee and the consequences of dishonesty, the General Security Programs are aimed at creating an appreciation and understanding of the Security Department's objectives as they relate to the specific industry they serve. Thus, in retailing, the whole mix of problems—including shoplifting, credit card frauds, hide-in burglars, counterfeit passers, quick-change artists, etc.—can be an interesting, informative and *educational* experience for employees, who walk away from such sessions with a deeper insight into the problems and with ideas as to what they can do in the future to prevent them.

Supervisory Training Sessions

New supervisors, while undergoing a new set of directions aimed at assisting them in their new responsibilities as supervisors, should be exposed to security problems that are peculiar to supervisors. What can and what should supervisors do under certain circumstances? What are their limitations? What are the company's expectations of supervisors under a variety of security conditions, such as the discovery of a break-in or major loss?

Again, as in the programs listed above, the best trainer is a security professional.

Employee Self-Protection Programs

Perhaps the most dramatic and best-attended employee self-protection programs are rape prevention sessions, using one of the fine quality 16mm motion picture commercial films available today. Employees are impressed that the Security Department is concerned about the protection of female employees as well as more business-related security activities.

Other employee self-protection programs, such as kidnaping prevention for executives, protection of personal property and home for regular (non-executive) employees, and basic self-defense are all possible programs the Security Department could offer, even on an optional basis, to employees of the company.

Again, this type of educational service is demonstrative of a security organization that cares about the company's employees. Consequently, the service tends to build a foundation of respect and support for the department's main objectives of protecting the company.

Unit or Departmental Presentations

Another important educational service role Security plays is in giving security presentations to various company units or departments. If a particular company unit—regardless of its organizational function or composition—wishes to hear from the Security Department, then the department should respond with a message aimed at that particular group. Housekeeping, Engineering, Purchasing, the Faculty Club, Merchandising Managers, the Youth Council—any group within the work environment is worthy of the Security Department's time and attention. (Sometimes it is necessary to cultivate an interest in security among the company's departments.)

The objective of each presentation, regardless of who the audience is, is twofold: first, to educate the group on the role and importance of the security function in the whole organization. This should be done in an entertaining and exciting way; the description of the security organization and its assignments can be liberally sprinkled with actual "war stories" which fascinate those not connected with the world of security. The second part of the objective is to point out to whatever group is being addressed how its role, contribution, or responsibility ties in with the security and protective efforts of the company or institution. In that way, they can identify with and relate to the security organization.

The educational efforts all strive to bridge the gap between the Security Department and the rest of the organization. For too long the gap has been an accepted fact; indeed, it has served as an insulation for Security. Unfortunately, that insulation has bred distrust and fear of the security function—a function that must, if it is to be truly effective, have the trust and support of all employees of the organization.

MANAGEMENT SERVICES

For the Security Department to make the maximum contribution to the organizational goals, security personnel (particularly at the managerial level) should achieve visibility as company management representatives *as well as* security management representatives. Specialists, as important as they may be, make limited contributions. Those who demonstrate interest in company problems and affairs, and who serve on various committees not specifically formed for pure protection purposes, play an additional, new role in the organization. They provide the company with a managerial support or service always in demand in organizational life.

This new dimension in Security's role must be sought out and cultivated, because Security has traditionally been content to limit its activities to the protective function. Organizational management, as a consequence, is accustomed to looking beyond the Security Department for general problem-solving counsel and assistance.

Summary

Security is primarily a protective service of prevention, most conspicuously engaged in such general protective activities as access control, cargo protection, building security, investigation of criminal activities, inspections, and enforcement of company rules.

Security can and should also provide many related services. *Special services* might include executive protection, bodyguard service, special investigations and emergency services. Security should be actively engaged in *educational services,* bringing security awareness to new and established employees, and to supervisors whose responsibility must include loss prevention. And, wherever possible, the effective security department will seek out ways to expand its role, making its presence felt as a general problem-solving arm of management.

Review Questions

1. Explain the statement, ''The value of the Security Department's service is better measured by what does *not* happen than by what does.''
2. Briefly stated, what is the general purpose of the security function in any organization?
3. Give four examples of special services which the Security Department might provide company management.
4. Describe how the Security Department might set up and operate an emergency service for the benefit of company management.
5. What are two objectives of the Security Department in making presentations to other company units or departments?

Chapter 4

The Director's Role

Definitions of titles in the world of security in the private sector are not as clear as those in the public sector. There is little confusion over the position of the Chief of Police within the Police Department or his status within the municipal government. On the other hand, the private sector tends to be rather indiscriminate in the use of the title "Security Director." Too frequently the "Security Director" is, in fact, a Security Manager. There is a difference between the two. The easiest way to differentiate between them is to consider to whom they report. A "Director" is ranked at the highest level of middle management and should report to a member of senior management such as the President or a Vice President. If the head of security reports to a lower level of management in the company's organizational structure, he is a manager.

The effective Security Director should have a track record of success in handling people and problems. He will be a dynamic, results-oriented individual with a high level of personal integrity. He should have the ability to develop organizational plans, to evaluate personnel and their assignments, and to supply direction (including new approaches where necessary) of the security function.

While all of the Director's activities come under the single umbrella of "management," it is possible to examine each of his important roles individually. He is . . .

41

    . . . a leader.

    . . . a company executive.

    . . . an executive with high visibility.

    . . . an executive with a broad profile.

    . . . an innovator.

    . . . a counselor and adviser.

    . . . a trainer.

    . . . a contemporary professional.

    . . . a goal setter.

The Security Director As A Leader

The Security Director's role as a leader means that he provides leadership to the management of the security organization. Note that he *does not directly manage* the department; he *provides leadership* for the manager and management team. Providing leadership means setting the right climate, pointing out directions, suggesting alternatives. The Security Director might be likened to the motion picture director. A fine film is a reflection of the director's talent, but he is rarely, if ever, in the film. So it is with the Security Director. He brings out the best of his people's talent and *they* perform.

The most difficult aspect of the leadership role is to refrain from making operating decisions. This is where the delicate art of good management skills comes to the fore. If the Security Director has selected and developed personnel properly, if he has given them *real* responsibility (and they understand this), if he has established a climate of confidence and professionalism, if he has motivated them —then his direction and suggested alternatives will allow his subordinates the courage, wisdom and strength to make decisions. And the Director must have the courage, wisdom and strength to let subordinates make mistakes.

As A Company Executive

The Security Director's role as a company executive means that he identifies with and is accepted by senior and middle management as part of the company's management team. He should not be viewed narrowly as a security man, but rather as a skilled executive (first) in the security field (second). He should not have the reputation or image of simply being the company policeman.

The Director's demeanor, deportment, grooming and attire should be equivalent to that of his peers.

As An Executive With High Visibility

High visibility means just that: a Security Director who is well known in and out of the company and who is seen frequently. Ideally the Director should be an interesting and effective speaker who is sought after to make presentations. The advantages of a popular Security Director over an unpopular Director should be obvious in terms of creating good will toward the security organization and its objectives.

Additionally, the Director should be visible—and available—to all the security ranks. He should make every effort to meet new security personnel, irrespective of their assignments, and seize every opportunity to chat with security people. That kind of visibility pays off in terms of employee motivation.

As An Executive With A Broad Profile

A broad profile means that the Security Director has interests in and contributes to other areas of the business beyond the security function. Such exposure and activity not only enhance his executive image but have other rewards as well. One benefit is that the Security Director has the opportunity to meet, talk to and work with people in the company whom he might never meet otherwise. Conversely, these people have the opportunity to meet and exchange ideas with the Security Director. The experience can be very rewarding and positive—good for them and good for the company (let alone the Security Department).

As an example of the Security Director's involvement in other areas of the business beyond security, in one large company the Security Director takes part in two different activities: College Campus Recruiting for company management trainees, and the company's Supervisory Training School for first and second level supervisors.

The Security Director's responsibilities have nothing to do with campus recruiting, and vice versa. However, the company's approach to recruiting is to utilize interested and qualified middle management personnel as campus recruiters. This brings to the recruiting effort a diversified range of experience and talents, functioning within the selection guidelines designed by the personnel executive who is an expert in campus recruiting.

As a result of this approach to recruiting, there are young men and women moving up throughout the company today who were initially selected by the Security Director. What do you imagine their respective attitudes are about security people in general and the Security Director in particular? As time goes by these people will move into ever more important levels of responsibility—and security needs friends, the more the better.

At the Supervisory Training School, the Security Director lectures on the subject of discipline and the disciplinary process. Attendance at the Supervisory Training School, a three-day program which must rank among the top in-house supervisory training programs in the industry, is highly coveted throughout the company. The program is meaningful and inspiring. Attitudes are changed. Skills are learned. Concepts open minds and eyes. In most cases the students are grateful for the experience and grateful to the lecturers who gave them that experience.

What is the beneficial consequence? The company has forty-nine separate facilities located in five states, and this Security Director cannot go to a facility without former "students" waving or coming up to greet him. These relationships foster greater acceptance and recognition of the Security Department's worth.

As An Innovator

The Security Director is constantly charged with the responsibility of finding new ways to do the job—better, less expensive ways —and thus he must be an innovative, flexible administrator. The term "creative security" is apt because the very phrase sparks one's imagination. "Is there a better way?" should be the Director's continual question. Innovation means experimentation and risk. Security tasks do tend to become entrenched, routine and *safe*, tried and true. There is a tendency to discover a successful formula for solving a problem and then stick with it. The trap lies in clinging to what succeeded for a period of time and resisting change, refusing to try a new formula because it is unknown.

An excellent example of an innovative approach to a communications problem in a widely dispersed security organization was the adoption of a telephone pre-recorded message system. Each day, every agent dialed the appropriate number at his appointed time to receive the Security Message of the Day; for example, the Salt Lake City agent called each day at 11:15 a.m. Rocky Mountain Time. The messages might be *instructional* (explaining a company procedure and how security could inspect for compliance), *informative* (reporting results of the department's activities, such as a major forger arrested), *warning* ("Be alert for counterfeit money orders bearing numbers L21344566 through 5699, with greasy feel around the indicia"), or *motivational* ("Keep up the outstanding work; all inspections were received prior to report deadline").

The idea of this daily communication tool solved a long-standing problem of disseminating information to the entire organization in a timely manner. Formerly, especially if the information was urgent, headquarters personnel would undertake the long and tedious task of trying to reach all members of the department by telephone. When agents were not in, messages left for them tended to become garbled and misunderstood. The innovative program not only transmitted the message to everyone each day, but also insured that all received the *same* message.

Another creative approach to a security problem in one retail organization was the shift away from total emphasis on theft detection to a rigorous loss prevention program. This shift was in answer to the staggering problem of an unacceptable inventory shrinkage figure. More arrests simply were not the answer to reducing losses. There had to be a better way than the traditional store detective and investigator approach.

Thus, because of innovative leadership at the Director's level, this company's "Red Coat" security program was born—a retail security program aimed at preventing shoplifting and other thefts instead of detecting them after the fact. Highly visible security personnel, dressed conspicuously in bright red blazers with gold emblems, have the job of discouraging, deterring, and preventing theft. If an act of theft is in progress or just completed, they attempt to "burn" it out—to discourage the thief by making him aware his conduct has been observed and he is under surveillance. If "burning" does not work, *then, and only then,* is an arrest made.

This program has finally balanced out to a remarkably successful ratio of 25:10 Prevention:Apprehension. It took courage to launch such a radically different approach to retail security in the face of long-standing tradition. But there *was* a better way!

As A Counselor and Adviser

Because of his wisdom and years of experience, the Director's role as counselor and adviser is an invaluable one to the department as well as to the company.

It is interesting to point out how frequently security management seeks the Director's advice on routine operational problems. More often than not, they ask advice in order to "test" or compare their solutions against the Director's. This is good, as long as the Director does not succumb to the temptation of grabbing the reins and *requiring* management to come to him or to include him in the problem-solving and decision-making.

His role is to give advice, suggest alternatives, *help* solve problems—not to solve them. He gives the benefit of his experience and judgment to the Security Manager and his staff. Occasionally, when a particularly difficult problem is under discussion and no answer has been developed for comparison with the Director's, the Director may hit on a solution which is immediately recognized by the Manager and his staff as *the* solution and consequently is adopted. In this kind of situation, the Director did not force his will on the subordinates; the climate was one of mutual and open exchange. The Director's involvement was participatory in nature.

Of course, the Director *can* make the decisions and solve all the problems. Some do, or at least approve each and every decision. But once the Director does this, he steps down into the role of his subordinate (the Manager under the Director in a large department). He is no longer *directing;* he is now actively involved in operations.

On the other hand, the Director sometimes will be called upon to solve a problem. In these circumstances he will interject himself into the decision-making process and force his will if necessary. But these situations, especially the latter, should be rare.

The Director is also an adviser to company management in terms of policy, construction planning, special events, emergency and disaster planning, executive protection suggestions, executive problems (such as an executive's daughter running away from home), and a host of other areas wherein the Director's good counsel is sought.

As A Trainer

The Director's attitude about the importance of the training and development of *every* security employee sets the climate for the department. If he is supportive of an aggressive, structured training program within his organization, that is what he will get. If he is lukewarm about training and feels that it takes away time that is necessary to get the job done, he will end up with a fragmented, ineffective program, if any. The Director's role as a trainer deserves as much stature as his other roles. It is certainly the one role that has an impact down through every level of his department, with the obvious end result of improved performance.

With respect to the organization generally, the Security Director's role as trainer is primarily one of a climate-setter. With respect to the staff, particularly the Security Manager or Assistant Director, however, his role is very functional. The Director must personally train, guide and develop his immediate subordinate, with the objective of preparing that manager to take over the Directorship at the earliest possible date. One reason is that there is no one else who can do it. Second, there is a moral responsibility to the subordinate to help him grow vertically. Third, there is a moral responsibility to the company to develop talent that can function in the Director's absence. Finally, effective management dictates that a replacement be ready so that the Director can move vertically to assume more responsibility—e.g., Corporate Security Director, Vice President of Loss Prevention, same rank but with a larger division within the corporation, or a more advanced position with another company.

The training of the manager does not terminate at some fixed point in time. It is ongoing in nature and more frequently than not it lasts for several years. Every day the manager is prepared to take over. The next day he is better prepared, and so it goes.

As he must develop and train his own people, so must the Director contribute to the training and education of all company employees as it relates to security and loss prevention. His input with the Training Department on induction programs for new employees, general security or loss prevention awareness programs, and special campaigns or promotions can make the difference between a very credible production and a program that is flat and ineffective.

As A Contemporary Professional

Being a contemporary professional means that the Director keeps abreast of the security industry—familiar with current case law affecting the industry, new and improved technology and systems, current trends, and the general state of the art. To accomplish this the Director must subscribe to and read trade journals, participate in local, regional or national security associations, attend seminars to hear his peers and see new products, and freely communicate and exchange ideas with contemporaries on a regular basis.

The importance of this professional role is better understood when one recognizes that the subordinate security manager and security technician are normally absorbed in the operating demands of their jobs and may be less free than the Director to peruse the vast array of information and data pertinent to the business.

The contemporary professional is constantly involved in training and educational programs. How can one be considered professional unless he is growing in his selected profession? This growth comes from broadened experiences coupled with new concepts, strategies, and tactics made known through some form of institutionalized educational process. Education in the security industry is not limited to the novice. Many security training and educational programs are specifically designed for experienced practitioners, supervisors and managers.

As A Goal Setter

Establishing objectives and setting goals for the organization is an important aspect of the Director's job. Who else could do it? If security goals are set by senior management there is no need for a Director. A subordinate cannot establish objectives to be met and tell the Director the strategy to achieve those objectives.

Goals obviously set directions, provide challenge, and should require genuine effort to be achieved. Goals which are too easily achieved are not real goals. For example, if one departmental goal for the coming year is to have one hundred percent of the department's supervisors graduate from the company's Supervisory Training School, and only ten percent (representing two or three people) have yet to go when that goal is set, then this is not a real goal. It is simply one of many things to be done on an ongoing basis. A goal must be an objective, an accomplishment that you have to shoot for, that you must work at constantly to achieve.

Goals, which must be quantitative or qualitative in nature, could include replacing personnel with hardware to reduce payroll dollars; converting a predetermined number of units to a new access control program within a specified time frame; reducing specific losses by a set percentage; improving a certain measurable skill of security personnel such as firing range scores; or designing and implementing a new Programmed Learning program for major disasters—to name but a few.

In addition to the major roles described above, the Director should wear a number of other hats that can be significant. He may be the departmental "Court of Last Appeal," "father confessor," listener, financier, departmental defender on a white horse, taskmaster, politician, professor and intelligence expert. He must be purer than Caesar's wife, and, finally, he must be a gentleman.

THE NEW SECURITY DIRECTOR

To be appointed the new head of security in a long-established organization, even if coming up through the ranks; or to come into an established organization from outside the company; or to be transferred from another area of the business to head up a newly created security organization; or to arrive on the company scene from outside for the purpose of setting up a program—all, to say the last, are difficult situations indeed. The new head of security is unknown and unproven (in *that* position), and everyone is suspicious of the unknown.

How can this natural suspicion of "the new man" be overcome? The answer is for the new security director to come in with the lowest possible profile. He should look and listen, and speak when spoken to, except when asking necessary questions. He should have a pleasant manner and concern himself initially with the people in his pyramid. Such concern must be sincere and warm. In private chats with each of them he will learn much without going out to seek it.

The new security director should be very conservative in terms of making changes, unless such change is badly and conspicuously needed. In that case, he should allow the change to be made, but not in his own name. He should allow the credit to go to a subordinate. People will suspect that the new manager is behind the change anyhow and quietly admire his style. The new manager should not threaten to "clean house," make sweeping changes, bring in "qualified" help or in any way forecast change; to do so tenses up the organization and prolongs the period needed for his assimilation into the environment. It is never wise for the new director to criticize his predecessor, if there was one. If criticism is due, it will naturally come from below. The new man should listen to the criticism and be prudent in his responses. A neutral response is best; then he should move the conversation on to positive statements about the future.

If a problem or question arises to which the new security manager does not know the solution or answer, he should say so. Just because he is the chief does not mean he knows everything. He should ask subordinates their advice.

This low-keyed, low profile, non-threatening approach—which even helps take some of the butterflies out of the new manager's stomach because he is not trying to prove anything—will buy time, and time is his ally. Changes will occur, of course, because the new man is there to insure protection for the company, and that means his style, his philosophy and his strategy will come into play with the passage of time. Loud noises and quick movements not only frighten animals and infants; they frighten adults, too.

Summary

The Security Director is commonly one who reports to a member of senior management; the head of Security reporting to someone at a lower level is more properly called a Manager.

Within the Security Department the Director's role is that of *leading* rather than operational decision-making. (The Manager of a smaller department will inevitably have more direct involvement in operations.) In his leadership role, the mark of a good Director or Manager is the ability to delegate responsibility and commensurate authority.

Outside of his own department, the effective Security Director should be a highly visible company executive, a part of the management team with interests that go beyond security. And in his relationship with his staff he will be an innovator, counselor, trainer and goal setter.

The Security Director moving into a new company or position will advisedly seek a lower profile initially than the one described above. He will seek not to force events and people, but to lead with patience and example.

Review Questions

1. What is the distinction between a Security Director and a Security Manager?
2. Give an example of how the Security Director may be involved in other areas of the company beyond security.
3. What are the reasons why the Security Director should prepare his subordinate to take over the responsibilities of the Security Director?
4. List three ways the Security Director can keep abreast of developments in the security industry.

Chapter 5

The Security Supervisor's Role

Supervision is comprised of many facets, including but not limited to hiring, training, discipline, motivating, promoting and communicating. Each of these factors is a specific skill unto itself. Rather than grouping all of these skills under the single heading of "Supervision" or "The Security Supervisor's Role," each will be examined individually. This chapter will deal with the supervisor and his relationships with those higher and lower in the organizational structure, his responsibilities, and general principles of supervision itself. Subsequent chapters will be concerned with those factors intrinsic to supervision.

In the smaller department the Security Manager may himself be directly involved in supervision, and the comments in this chapter on the supervisor's role would obviously apply to the manager.

One popular definition of supervision is the task of getting others (subordinates) to get the job done, the way management wants it done, when management wants it done—willingly. Willingness, of course, is the key aspect of this definition. We are interested in enlightened supervision, not the kind of supervision used to oversee the slaves on the plantations of years ago, or the slave labor programs of Hitler's regime, or even Soviet operations of today. Historically, such autocratic methods, by and large, do get the job done—but not always at the time or in the manner desired. In a free society the most difficult part of the supervisor's task is to get the job done willingly.

The supervisor's job, then, is to get other people to accomplish tasks, which means they must perform. *Performance* is the ultimate responsibility and goal of supervision. Everything revolves around job performance—execution at the line level. The supervisor's performance (his supervisory skills) is reflected in the performance of those who work for him.

The Supervisor As An Inspector

There is an old adage that says, "Employees don't do what you expect, they do what you inspect." More often than not, that is true; not because they do not want to or do not care to perform their tasks, but simply because of human frailty. That same element of human failure is not limited to line employees; it can be traced to every level of every organizational structure, right to the top. From the top down, therefore, each "supervisor" must inspect the work of his subordinates. The Director inspects the Security Manager, the Manager his middle managers, the middle managers their supervisors, and the supervisors their subordinates. When that inspection process breaks down, for whatever reason, tasks break down, deadlines are missed, other tasks are temporarily neglected and eventually forgotten. It is a source of amazement to all levels of management that functions, tasks, duties, reports—all assumed to be taking place with regularity—have "slipped through the cracks" and disappeared from organizational life, all because the inspection process failed.

On the other side of the coin, inadequate inspection frequently surfaces when a change in supervision reveals tasks or reports being religiously accomplished which no longer serve their original purpose. Often, no one seems to know who started the tasks, or what they were intended to accomplish.

The inspection need not, and should not, be a negative process wherein the supervisor tries to find errors or omissions, then criticizes. That managerial style creates a climate of resentment, defensiveness, and hostility. One can *always* find fault.

The most effective managerial style in the inspection process is to find those tasks which are done properly, acknowledge and give credit for good performance in such areas, and then point out deficiencies in an objective fashion. Most employees want to do a good job. Most failures, as already indicated, are the result of human frailties and not of malicious design. Consequently, when performance deficiencies are pointed out objectively, they are usually received with some embarrassment, on one hand, and an expression of genuine desire to improve on the other hand.

To be effective, this critical process of performance inspection must be consistent, continuous, constructive, and tailored to the individual employee.

The Supervisor and the Individual Employee

The supervisor must deal on an individual basis with each subordinate, because every employee is different. Every human being on the face of the earth is different. The differences are manifested not only in observable physical features and fingerprints, but in how each individual responds to external stimuli, how he perceives things, his beliefs, fears, aspirations and needs. Such human differences mean that different people require different handling. Some may require more supervision than others. Some respond to persuasion, some to command. Some want to set goals, some want goals set for them. Some are uncomfortable around authority figures, some are at ease. Sensitivity to employee differences is one characteristic of a good supervisor.

Supervisory Authority

A supervisor must have commensurate authority to carry out his responsibilities. If a supervisor is told he has the responsibility of ten security officers to protect the facility between 4:00 p.m. and midnight, and at the same time he is told that any disciplinary action against any one of those ten will be handled by the next level of supervision, then he has been denied the necessary tools or stripped of the necessary authority to carry out his responsibility. Such conditions, which do indeed exist, make a mockery of organizational integrity and turn what should be legitimate supervisors into "straw bosses" or lackeys. The supervisor represents management and must be given the necessary authority to make that representation meaningful. If, for any reason, appropriate authority cannot be vested in a supervisor, it will still be necessary to have some form of "lead man" in charge.

The supervisor—with his officers, agents, investigators or whatever their titles—should have not only the necessary authority to discipline, but he should also have some input in the selection or assignments to his unit. He should be heard when his people are considered for promotion; he must have authority to require additional training; he must have the authority to communicate to his people, including sending instructions, memos, etc.; and he must have the freedom to measure his people's performance without interference.

The issue of a supervisor's need to measure his people's performance without interference is even larger than the one of disciplinary rights. Here is a typical case: The supervisor is obliged to evaluate the performance of his subordinates on an annual or semi-annual basis. He follows directions in terms of completing the personnel form prepared for each employee. (A sample Employee Evaluation form is shown in Appendix A.) He marks the various boxes that represent rating factors such as "submits reports in a timely manner," and finally makes an overall evaluation as "Above Standards." He submits his evaluations.

Two weeks later he is called before his superior or the personnel office and is advised that his rating of "Above Standards" of Officer X is too high. The rating should be "Meets Standards." Despite the fact there are no flaws or inconsistencies in the various factor-ratings with the overall rating, the supervisor is instructed, for whatever reason, to reduce the rating to the next lower. Or, for that matter, he is instructed to *raise* the overall rating to the next higher.

The above problem is frequently a point of concern in any discussion of supervisory training and practice. Those raising the question say, "I don't want to get into trouble with anyone, so I changed the rating, even though I felt my evaluation was correct. What should I have done, or what should I do next time?"

For the responsible supervisor, the answer is clear. If he is convinced that his evaluation is not incompatible with the firm's definition of standards, and the evaluation is not inconsistent with his other evaluations, then he should seek to support his own rating. If, after sufficient discussion, he is still asked to change that rating, he may have no alternative but to make the change as requested. In such circumstances, he should indicate on the evaluation form that the rating is not his own.

If a supervisor is not capable of disciplining, then he should not be a supervisor. By the same token, if he is incompetent to evaluate the performance of subordinates, he should not be a supervisor. If it is a question of skills in disciplining or evaluating performance, the onus rests with management to provide adequate training to develop such skills—not to take that authority away.

The "In-Between" Man

The supervisor is the vital link between the employee and security management. The supervisor represents management's needs and views to those below and at the same time has the responsibility of representing the needs and views of his people up to his management. Failure to discharge this function objectively and faithfully, in a timely manner, can have disastrous results. The supervisor who, being closest to the scene, is aware of sentiments, grievances or problems, yet does not inform management, fails twofold. First, he fails his subordinates by not carrying the message to management. The condition, whatever it may be, is allowed to deteriorate, to the disservice of his subordinates in terms of morale, accidents, or turnover, depending upon the problem. Second, he fails management by withholding information that could provide them with answers, explanations, or decisions to resolve the issue.

This intermediary status is usually well understood by line personnel. It makes sense. Yet that status can easily serve as a crutch for the weak supervisor, providing an excuse to shirk responsibility so that all distasteful duties or assignments, or decisions which may be unpopular, are passed off (even if they are his own) with the disclaimer, "Management wants it this way." The weakness is apparent: this supervisor wants to be popular all the time. Fortunately, that kind of supervisory weakness cannot be concealed for too long.

Span of Control

Span of control, the number of employees a supervisor can manage, depends upon a number of factors. One important factor is the supervisor himself—his skill level in handling people and his ability to delegate responsibility. Another factor is the job description of his subordinates. Field investigators with relatively sophisticated assignments require more attention from the supervisor than a uniformed staff assigned to one location on one shift. In the former case, the proper span of control might be six and in the latter, twelve. Long-standing and widely accepted span of control standards suggest the following ratios:

	Supervisors	:	Employees
Ideal	1	:	3
Good	1	:	6
Acceptable	1	:	12

These numbers represent spans of control under normal operating conditions on an ongoing basis. However, under certain circumstances, for a relatively short period of time and with a homogeneous group, one leader could handle up to two dozen employees.

One Boss

The Principle of Unity of Command is the classic or traditional way of saying that every employee must report to only one superior. Find a situation where a person is being directed by more than one superior, and you will find that subordinate coping with conflicting instructions and confusion, resulting in diluted performance.

Consider the frustrations experienced in one actual situation by the Chief of Campus Police for a group of adjacent private colleges in Southern California. In that position the Chief was responsible to five college presidents, each of whom had his own particular point of view. In one incident, a group of students had gathered off campus in a neighboring county for a Friday afternoon T.G.I.F (Thank God It's Friday) party. The Sheriff's Department of that county took this large group of students into custody for possession and consumption of alcoholic beverages. One student slipped away, advised the Chief of Campus Police of the events, and stated that the Sheriff's officers were calling in buses to transport the students to the county jail.

At the scene the Campus Chief discussed the matter with the

officer in charge and convinced him the best interests of justice would be served if he would release the students to the Chief, who in turn would process them through their respective college student court systems. The Sheriff's office could see the wisdom of avoiding the booking hassles and subsequent difficulties of proving in the county court just which student was doing what (there were close to 100 students).

The Chief of Campus Police escorted all the students back to his office and had them line up for identification purposes. He then submitted lists to three different college student courts. The courts, as expected, levied substantial fines and built up the coffers of the student body fund, and justice prevailed—at least the Chief thought so.

The outcome of all this was that one college president expressed warm appreciation for the Chief's intervention, which had saved the school from what would have been certain unfavorable publicity at the hands of the local press, and most of all for the avoidance of criminal booking records for his students. However, one of the other college presidents took exception to the Chief's intervention. His position was that the students sooner or later had to assume responsibility for their conduct. They had been warned about assembling at that particular location for "beer busts" before, and therefore they should have experienced the full consequences of their conduct.

The point is, this security director could have lived with either position had he worked for either president, but he worked for both. His job was to serve both, and obviously he could not please both.

The employee, then, who has more than one supervisor can find himself in an unworkable situation. The organization must follow the principle of Unity of Command to avoid such counterproductive conditions.

Automatic Shifting in the Line of Command

There are necessary and legitimate exceptions to the principle of Unity of Command. Conditions that require another supervisor are:

1. Emergencies.
2. When the failure of a ranking employee to take command would jeopardize the department's objectives or reputation.

An example of Number 1: A uniformed guard, immediately following a natural disaster such as an earthquake, is approached by a security investigator (who is in an entirely different departmental pyramid or line of command but has rank over the guard). The investigator instructs the guard to run to the side of the building and cut off the gas supply. The guard cannot refuse this shifting in supervision.

An example of Number 2: A uniformed guard on a parking control assignment for a major event has been instructed by his supervisor to deny access to one reserved parking lot. A supervisor other than the guard's, because of his mobility and overview of the parking and traffic conditions, reaches the opinion that the congestion can only be relieved by routing traffic into the empty lot. Not to relieve the growing congestion could have serious repercussions on the event itself. The supervisor, knowing he is accountable for his decision, can command the guard to let the cars in.

Such direct orders out of the normal chain of command are invariably given under a time pressure; that is, a decision and action must be immediate. The consequences of delaying action in order to locate the proper supervisor could be serious, if not grave.

Such automatic shifting in the line of command, always of a short duration, requires full understanding on the part of all department members at all levels. It is interesting to note that such shifting *does not* violate the principle of Unity of Command; rather, it enhances and supports the principle by having a rule and understanding of the exception. Exceptions add credence to rules.

Functional or Staff Supervision

Although every employee has his own supervisor, there are numerous occasions and conditions where the employee must perform at a time or location outside the immediate control of his supervisor. An example would be an alarm operator and alarm serviceman working the graveyard shift. Their supervisor works the day shift. By agreement, the graveyard watch commander, in another pyramidal structure within Security, assumes functional (or staff) supervision over these two security employees. As a functional supervisor he has responsibility for a limited degree of supervision, but not complete control.

There are two aspects to this functional supervision. The first is that the watch commander in all probability has no technical competence in alarm operations or servicing, so he cannot give commands which would interfere with performance. This means his supervision is limited to such things as promptness, following general orders applicable to all personnel, and demeanor on the job.

The second aspect of functional supervision is that it is essentially advisory in nature. The functional supervisor can discuss problems with the subordinate, make suggestions, point out mistakes, but—and this is most important—he lacks authority to take disciplinary action. Certainly he can cause disciplinary action by reporting the problem to the subordinate's supervisor; that supervisor can take the corrective action, but the staff supervisor cannot.

Supervisory Training

Perhaps the most common shortcoming in the security industry is the failure to properly prepare and equip new supervisors with the tools to discharge their important responsibilities. A line employee on Friday may become a new supervisor on Monday, with no distinguishable difference in the eyes of former peers.

It is better to give the new supervisor a week off with pay and have him sit in the library studying books on the fundamentals of supervision than to put him in his new assignment immediately. If he is off for a week and the employees believe he is attending some special training just for supervisors, they see him, upon his return, with entirely different eyes.

The library trip, of course, is a barely acceptable alternative in the absence of what should really happen. Ideally, the new supervisor, before he takes command, should attend a workshop, seminar or training session for new supervisors. If an in-house program is not available, he should be sent to a commercially conducted program, of which there are many to choose from in most communities.

The problem in this area is the false assumption that, because a man was an outstanding investigator or officer, he will make a good supervisor. That simply does not follow automatically. A new supervisor must master an entire new set of skills that have absolutely nothing to do with investigative ability. Such skills must be learned; they are not inherited and do not come into play automatically upon promotion. And new supervisors are quick to discover that they are ill-prepared for their new responsibilities. They are sensitive to their deficiencies and lack confidence in handling problems and people. Subordinates are very quick to sense this absence of confidence, and some will not be at all sympathetic but will capitalize on the apparent weakness to their own advantage—especially those who are of the opinion *they* should have received that promotion.

If, for a variety of reasons, it is impossible to send the supervisor to a school immediately, or to the library, the next best thing is to arrange for his attendance at a later date. Subordinates, knowing the new boss will indeed be attending special training in the near future, will respond in a more supportive posture than in the complete absence of training.

Summary

Supervision has been defined as the task of getting work done how and when management wants it done—willingly. *Performance* is the ultimate goal of supervision.

The effective supervisor will best assure the performance of his subordinates by constant *inspection*. He will be sensitive to individual employee differences in providing both criticism and encouragement. He must have the *authority* to carry out his responsibilities, both in discipline and in employee evaluations. And as the man-in-the-middle between management and the employee, the supervisor should play an active part in the process of communication both upward and downward.

Important principles of effective supervision are *limited span of control* and *unity of command*. The latter principle (no man can serve two bosses) is not violated by *functional supervision*, which is exercised over employees only temporarily under a supervisor's control. Functional supervision is both limited and advisory in nature.

A good employee does not necessarily make a good supervisor. Effective security management will provide adequate *training* for new supervisors. Training inspires confidence in both the supervisor and those who will serve under him.

Review Questions

1. Define *supervision*. What is the ultimate responsibility and goal of supervision?
2. Explain how you would handle the situation if you were asked by a superior to change your rating of a subordinate's performance.
3. In what way is the supervisor "the vital link between the employee and security management"?
4. What are two factors influencing a supervisor's effective span of control?
5. Give two examples of legitimate exceptions to the principle of Unity of Command.

Chapter 6

The Individual Security Employee

Relatively little instructive material is available concerning the individual security employee's role, contribution or importance in the overall security function. The truth of the matter is, the good reputation of the Security Department and the successful achievement of the department's objectives are, ultimately, the result of the employee's execution of his job. Poor performance, poor reputation. Good performance, good reputation. Excellent performance, excellent reputation.

Regardless of how knowledgeable security management may be, the line employee's performance is the measurement of success. This performance has two dimensions: application of skills, and general conduct. Specific skill development, skill levels and execution of skills are the subject of many texts and will not be our concern here. Rather, we will examine the issue of conduct, not only because it has been touched on so lightly within the industry, but because, in its critical importance, conduct frequently transcends skills. In other words, if a security administrator had to choose between good skills but poor conduct and poor skills but good conduct, more often than not he would choose the latter.

Conduct, then, plays a significant role in the Security Department's general reputation. The company which Security serves expects exemplary conduct of its security force. Security management must demand exemplary conduct. Line employees will respond to such expectations and demands if they know and understand what the standards of conduct are. It is a truism that most employees will do what management wants, *if they know what management wants.*

In the absence of any generally circulated or official standards of conduct in the security industry,* the following is submitted:

STANDARDS OF CONDUCT

1. Security employees are habitually courteous and attentive to those seeking assistance, reporting conditions, or lodging complaints.
2. Security employees are punctual and expeditious in the discharge of their duties.
3. Security employees conduct themselves in a just and objective manner, treating all with equal reasonableness.
4. Security employees consistently exhibit a spirit of cooperation with all and do not allow personal feelings to interfere with their work.
5. Security employees conduct their personal and business life in an exemplary fashion, above reproach in terms of stability, fidelity and morality.
6. Security employees have a cheerful and positive approach to their work.

Today these standards may sound idealistic if not old-fashioned, but if they are adopted and followed, the end results will have a favorable impact on a department's reputation.

Courtesy

Courtesy starts at home, granted; but the development of courtesy on the job starts with mutual respect for fellow security employees. And one cannot be expected to respect his associates unless he respects himself first. Security management's responsibility in

*For additional discussion of this subject, see *Private Security: Report of the Task Force on Private Security* (Washington, D.C.: National Advisory Committee on Criminal Justice Standards and Goals, 1976), especially Standard 3.2, "Conduct of Security Personnel."

this area is to insure that the employee has dignity—dignity in terms of pride in his uniform, his work place, his personal responsibility in his work. Standards set in these areas have a definite influence on an employee's assessment of his own worth.

Issuing or permitting the use of shabby uniforms, for example, takes away from a man's sense of pride. High standards for uniforms, on the other hand, automatically instill self-pride and, hence, self-respect. Old and inadequate office equipment and furniture in poor condition have a demoralizing effect, whereas equipment and furniture in fine condition make employees feel valuable to the organization. And, of course, the knowledge that each day's work is important and contributes to the overall success of the department is most necessary. The trouble is that many employees do not understand or see that their daily contribution is significant, usually because management has not bothered to tell them.

When an employee works within a climate which fosters feelings of self-worth, he will normally feel that his co-workers are also important and worthy of respect. Respect and courtesy will radiate beyond the Security Department—*if they are an understood standard of conduct.*

The emphasis on "if they are an understood standard of conduct" is important because of the nature of the security business, particularly those types of security organizations that are heavily engaged in the detection and apprehension of criminal offenders. Frequent, regular contact with negative or antisocial people tends to harden the security officer, just as it tends to desensitize law enforcement officers in the public sector. Unless courtesy is demanded under all circumstances, it may be practiced within the organization but may not be demonstrated consistently outside the organization.

Of the six standards suggested, the one of courtesy is the most conspicuous and visible to the beholder. Courteous behavior is not restricted to personal contacts; telephone courtesy, or the absence of same, is also very visible. A great deal of security business is conducted over the phone, and employees in the security organization, including clerical workers, must understand the importance of telephone courtesy.

Everyone, including scoundrels, needs and likes to be treated with courtesy.

Responsibility

Standard number two speaks to "punctual and expeditious" performance. The individual who is not punctual is not a responsible person. To be late for duty, to be late with reports due, to be late with assignments, all reflect a lack of responsibility. Such a person is expressing the attitude, "I don't really care about what's happening." Children lack responsibility; maturity is a factor in assuming responsibility. Children are impatient and soon weary of details. Concern for details, including time, is another measure of responsibility and maturity.

Webster's New Collegiate Dictionary defines *expeditious* as "Possessed of, or characterized by, efficiency and rapidity in action. . . ."

For the department to have a reputation of being prompt and efficient, every member must be prompt and efficient. Much security work involves reporting facts (details). A lackadaisical approach, an absence of concern over details, and unnecessary delays are all serious reflections on an organization. Every security employee must be punctual, attentive and accurate. Company as well as security management relies on the security employee's sense of responsibility.

This may seem self-evident. It is a question of self-discipline, and the setting of standards imposes self-discipline.

Due Process

Standard three requires that security employees act "in a just and objective manner, treating all with equal reasonableness." Essentially this means respect for the rights of others. To be other than just or to be subjective, particularly in the enforcement aspects of security work, is to give more rights to some and to deny the rights of others. Such conduct is intolerable in the public sector, as evidenced by many appellate court decisions restricting law enforcement, as well as by the public outcry over the Watergate scandal. As of this writing, the courts have not yet reached down into the private sector with binding restrictions in terms of our enforcement activities, but that is not to say that they will not.

Actually, respect for the rights of others is a state of mind and should not require legislation. Security professionals can create the proper state of mind through standards of conduct expected of their employees.

The *due process* concept is very meaningful in an organization, such as a large retail store or any organization which takes scores of people into custody every month of the year. To be objective and treat each one with equal reasonableness is a challenge in view of many attempts at violence as well as verbal abuses. But it is both practical and humane to treat all suspects with respect; after all, once they have been apprehended the victory is already won. There is no need to harass or further embarrass them, to be verbally abusive or gloat over their misfortune in having been caught. Even in the face of vile verbal attacks, members of the staff should impassively and objectively go about their business of completing reports and related tasks with an air of quiet dignity which evokes nothing but respect from onlookers, from management on the scene, and from the police who arrive to assist.

A store detective with an abiding respect for the rights of others is a store detective with the smallest ratio of false arrests. Some may say the smallest false arrest ratio is purely skill-related; in truth, however, the best detective not only has masterful skills, he or she is also sharply tuned into the consequences of a questionable arrest—consequences not only to the store and the possible civil liabilities connected therewith, but also the mental trauma and anguish of the innocent person. They are sensitive to human rights. Insensitive detectives tend to be rash. They will gamble, will act on "instinct," will take the word of another rather than what they know through their own senses. Insensitive detectives, and all other classifications of security personnel, can become sensitive through adoption of standards of conduct.

Cooperation

A reputation of willingness to work with and for others, to serve and assist, materially contributes to the department's good image. Far too many departments attempt to find reasons not to do a job or reasons why they cannot. From the Security Director down to the newest employee with the most limited responsibilities, the watchword should be "Why, certainly"—assuming, of course, that the requested service is possible and not contrary to the best interests of the organization and there is no specific rule against it.

For example, the lobby desk officer may not have facilities to store briefcases. He could easily turn down a request from a visiting dignitary to watch his briefcase. But if he says, "Why, certainly," and places the case behind the desk for fifteen minutes, he creates a most favorable impression. It is the spirit of cooperation that counts. The dignitary would not ask unless he had a need, and he would not ask a particular person unless he felt that person was responsible. A response of "no" or "I can't" is difficult to accept.

The more Security can do as a service, the more important the entire operation becomes and the more company management will look to Security for such service. For example:

Executive: Could you spare a security man to pick up a visitor at the airport? The taxis are on strike, you know.

Security: Why, certainly.

Public Relations Officer: Could you have one of your men help out at the entrance to the special event? We need another pair of hands to collect the passes—we're afraid of gate-crashers.

Security: Why, certainly.

Employee (to security patrol in the parking lot): Could you call on your radio and ask for a tow truck to come out here? My starter won't turn over.

Security: Why, certainly.

Every member of the department should be coached to look for ways to serve instead of looking for ways not to. "Why, certainly" is a magic phrase. And it is attitudinal. Develop that spirit in the security organization, and the Department's reputation will be enhanced.

Personal Integrity

Standard of conduct number five states: "Security employees conduct their personal and business life in an exemplary fashion, above reproach in terms of stability, fidelity and morality." The terms "above reproach" and "exemplary fashion" are very broad but hint of such qualities as self-respect, honesty, cleanliness, fair play—qualities of universal appeal. At the risk of becoming trapped by moralistic values and definitions and the interpretations of same, let us consider some of these standards as they relate to the security officer's job.

Self-respect. As discussed earlier in this chapter, self-respect means dignity and pride in oneself and what one does. An individual's sense of pride and self-respect is mirrored in the way he walks, holds his head, looks others in the eye, and executes his assignments. A security employee who takes pride in his reputation and his work will perform in an exemplary fashion.

Honesty. Honesty here refers to the smallest corners of our character, areas such as the tendency to exaggerate. Honesty in words as well as deeds is required of every security professional at every level.

Security people are subjected to more temptations than perhaps any other career field—an "occupational hazard" of a sort. Only a strong conviction of what is right and what is wrong can provide the necessary strength to resist the temptations. Again, more often than not it is the little things that test men. Everyone has heard others say, "If I ever steal, it's got to be worth it—say five million!" The truth is that most of these people would not really have the courage to commit a substantial theft, but might take something small and unnoticeable, like a handful of salted nuts from a candy counter display, or a glass of root beer from the soda fountain in the employees' lounge.

Cleanliness. A person's personal habits can be pleasant and rewarding to others or can be loathsome and offensive. Most security people are highly visible, and their contact with other company employees or the public must be positive. People notice other people. Dirty fingernails, ear wax, nose hairs, body odors are all correctable.

Stability. Consistency in action and reaction is most important. The luxury of being moody cannot be permitted in the security organization. A moody security officer at the front lobby desk would not last long.

Excitability is another unacceptable characteristic in a security man's personal makeup. Being moody, excitable, having a tendency to lose one's temper—these types of mental peaks and valleys detract from exemplary conduct.

Fidelity. Fidelity means the careful observance of duties as well as loyalty. A loyal security employee is steadfast and true, dedicated to the organization. To complain about departmental policies, procedures, assignments or personnel is not disloyalty but lacks the true ring of fidelity. That is not to say, however, that a legitimately channeled and righteous complaint takes away from one's loyalty. There is a fine line between what one could call a positive complaint and a negative one. Perhaps the difference lies in the manner in which the complaint is levied. The writer can immediately call to mind a supervisor with years of faithful and loyal service whom he considered absolutely dedicated to the organization. Rarely did this man complain, but when he did, it was always very quietly stated, upward.

Dissatisfaction on the job does reflect in one's performance. This obviously suggests that poor or questionable job performance could be an indicator of unhappiness and dissatisfaction. Dissatisfaction erodes loyalty. It must be identified and dealt with as early as possible in a very up-front, open and honest manner. If the source of dissatisfaction cannot be corrected or explained satisfactorily and the employee is still not happy, he should be advised that the organization cannot change to accommodate him. He must adapt to the organization; if he cannot, he will remain dissatisfied. If he remains dissatisfied it will affect his performance and his work record. In view of that, it should be pointed out to him that under such circumstances the honest thing to do is to change organizations, for his best interests as well as the organization's.

Morality. The security officer should observe a standard of excellence in terms of right or proper conduct. Because of the high visibility of security personnel, employees in that capacity with questionable or low moral standards attract attention. Once a poor reputation is established, it is difficult, at best, to reverse.

Not only does the poor reputation of individual officers impinge on the department's reputation; the real threat or hazard, from the administrator's point of view, is compromise. Once just *one* security employee is compromised, the organizational objectives are compromised. To maneuver for compromise based on sexual behavior is a commonplace of intelligence, political and organized crime strategy today. To treat so-called sexual freedom lightly in the security context would be naive and counterproductive.

There are obvious limitations on security management's influence on employees' standards of morality. Management cannot dictate what movies they can or cannot see; what books they should or should not read; or what personal relationships would or would not be acceptable. What management can do is discourage improper conduct by (1) setting high standards; (2) insuring that every security employee knows those standards; (3) educating the employees as to the hazards of compromise; and (4) prohibiting fraternization with company employees. How effective those actions prove depends a great deal on the effectiveness of the organization's employee selection and screening processes.

Attitude

Of all the characteristics or virtues one brings to the job, none, including educational achievement, can exceed that of attitude. Attitude determines one's conduct and bearing toward others and their reaction to him. Good attitude produces good reaction; bad attitude, bad reaction. Attitude is contagious. It is a key ingredient in success or failure.

Take a security employee exemplifying the other five standards of conduct—one who is courteous, punctual, objective, cooperative, above reproach—imbue that employee with a cheerful and positive attitude, and you will have the finest security employee. And the department's reputation is a reflection of that composite.

Summary

The Security Department's performance ultimately depends on the performance of the individual security employee. And that performance is measured by conduct as well as by specific job skills.

Standards of conduct for the security employee should emphasize basic *courtesy* toward fellow employees and others, even in the handling of transgressors; *responsibility* in carrying out his duties; *fairness and objectivity*, respecting the rights of others; a spirit of *cooperation*; personal *integrity*, both on the job and in personal life; and a cheerful, positive *attitude*.

Security employees who exemplify these standards enhance both the reputation and the effectiveness of the security function.

Review Questions

1. What can management do to foster feelings of self-worth among security employees?
2. Explain the concept of *due process*.
3. What would you include in your own "Standard of Conduct" for security employees?

Part II

SECURITY PERSONNEL MANAGEMENT

Chapter 7

Hiring Security Personnel

Bringing new people into the organization is one of the most important functions, if not *the* most important, of the Security Department. After all, the department *is* people; their skill levels, intelligence, sense of well-being, grooming, morality, industriousness— all these and other factors dramatically affect the image and reputation, let alone the performance, of the Security Department. Investment in a careful and selective hiring program pays good dividends. Anything less is risky and has the potential of a swift negative reaction.

Hiring is a step-by-step process that eventually leads to the applicant's acceptance of a job offer. These steps are:
- Recruiting activity
- Interviewing
- Secondary interviewing activity
- Selection of best candidate
- Background investigation of candidate
- Job offer.

Recruiting

Entry or First Level Positions. A direct approach in advertising the fact that a vacancy exists is usually desirable. This openness can include information such as company name, the fact that it is an

Equal Opportunity Employer, location of job, uniform benefits (or requirements), starting salary, minimum requirements, and the fact that the position is an entry level job. As a rule the Personnel Department of the company administers the recruiting activity; however, they look to the individual managers for direction. *Within the limitations of company policy,* the Personnel Department strives to meet the manager's wishes. Policy restrictions such as "No salary quotations in newspaper advertising" would obviously have an impact, although not an adverse one, on the degree of openness in advertising.

The issue of advertising salary is quite controversial. The salary question must be answered at some point, and it will indeed be a factor in the applicant's decision. Just as the company is in the market for new employees, the applicant is shopping for a new employer. Based on certain data available in the newspaper ad (or whatever the medium may be), the applicant *selects* prospective employers. How many used automobile advertisements in the newspaper go unanswered because the seller withheld the price of the car?

In addition to newspaper ads, entry level security applicants may be solicited by posting announcements on bulletin boards in the security administration or criminal justice departments of local community colleges. If a college does not have a security or criminal justice program, the opening may be posted with the school's placement office. College students constitute a great reservoir of manpower for entry level positions—sometimes with, but more frequently without, career intentions. They are quick to learn and are usually willing to work those shifts or hours considered least desirable.

Non-Entry Level Recruiting. The recruiting approach for skilled, technical, and managerial personnel is different from that for entry level positions. Rather than the direct, open approach, the "blind ad" technique is recommended. Such advertisements are designed to attract career or professional people. Advertisements must appeal to and solicit their specific talents; for example, the copy might read, "Major banking firm's Security Department accepting applications for position of Fraud Investigator. Applicants must have minimum five years credit fraud and/or forgery investigative experience."

Some candidates reading such an ad will say to themselves, "That's me. I qualify." If they are in the market for a change or a new job, they will respond. The "blind ad" is simply one in which the company's identity is not revealed. Instead, interested parties are directed to submit their resumes to a post office box number or to some other third party.

The unidentified advertisement permits the company to pre-screen candidates and interview on a highly selective basis. It also allows some time for at least a preliminary background investigation into the candidate's qualifications prior to the initial interview. This is the key to the two opposing types of recruiting techniques. The entry level positions require rather broad, general qualifications that are possessed by a greater segment of the labor market. Such applicants select their employers. In the advanced positions in the department, on the other hand, the company is seeking specific candidates with specific skills. The organization knows exactly what it wants, and it will select the future employee.

Finally, skilled, technical or managerial candidates will be filling far more sensitive positions in the security organization than will entry level candidates. For this reason, far greater care must be exercised in the selection of advanced candidates.

Interviewing Preliminaries

An applicant's first contact with the company should be with the Personnel Department. Even though the applicant has submitted a resume of his background and experience, his application for employment with the company should be formalized and documented by the completion of the company's standard job application form.

Every applicant's first interview should be with a personnel interviewer who will review the data on the application, making any corrections and clarifications as appropriate. This initial personnel interview is not for the purpose of selection or making an employment decision. Rather, it is an official preliminary, preparing the applicant for the coming interview with the Security representative.

The applicant is then escorted or sent to the Security offices with his employment application, preferably sealed, for the real interview. (Many applicants look upon the personnel interview as a nuisance and are anxious to talk to the person they believe has the authority to make a hiring decision—the Security official.)

Following the interview, the Security representative will make notes on the reverse side of the application form concerning the decision he or she has made. The security interviewer must, however, be properly trained in those laws specifically pertaining to hiring practices, both at the federal and state level, that prohibit discrimination against applicants based on sex, age, race, creed or color. For example, the notation on a female candidate's application that "mother-in-law baby-sits" could be construed as sexually discriminatory should the applicant not get the job. (Would the interviewer ask a male applicant if he would have trouble getting to work because of baby-sitting problems?)

The question of discrimination becomes particularly important when an applicant is rejected. Any subsequent claims of discrimination in hiring practices will then be processed and administered by the Personnel Department. It is the Personnel Department that has the expertise and resources to handle such problems efficiently, not the Security Department. For this reason, it is foolhardy to bypass the Personnel Department and talk to possible candidates about employment opportunities privately.

The Interview

The purpose of the interview is for the interviewer to determine if there is a match between the interests and qualifications of the applicant and the needs of the department. This can only be achieved on a personal, one-on-one basis.

Prior to the commencement of the interview, the interviewer should study the written application in private. It is disconcerting for the applicant to sit in silence watching the interviewer pore over his application. Likewise the interviewer will find it difficult to concentrate on the application with the applicant staring at him. As he reviews the application, the interviewer should make a mental note of two or three highlights that he will want to explore in some depth during the interview. Throughout the questioning, the interviewer should feel free to refer to the application, but he should not make the too frequent error of repeating to the applicant the same data presented on the application.

For example:

Interviewer: I see that you worked for Bignam Glass Works for three years as an Investigator.

Applicant: Yes, sir.

That exchange does nothing to help determine whether the applicant offers the qualifications desired. It is better to give him a chance to provide that information in his own words.

For example:

Interviewer: Tell me a little about your experience at Bignam Glass Works.

Applicant: Okay. I joined them as a trainee when I graduated from State U., on a special projects assignment . . . mostly compiling statistics for department manpower and budget projections. Six months later an investigator position came up and I got the promotion. I was assigned to background investigations while I was there.

This exchange gives the interviewer some meaningful information to consider or explore further. For example, the next question might ask the applicant how much time it took to complete a typical background check.

Rule #1 in interviewing, then, is:

- *Ask open-ended questions that cannot be answered with a yes or no.*

Rule #2 is:

- *Do not signal the answers you are looking for in your question.*

For example, the question "Did you ever have to fill in for a supervisor and have people report to you?" is a signal to the applicant that the interviewer considers some supervisory experience very important. Naturally, the applicant will tell him what he wants to hear: "Oh, yes, sir, a number of times the Special Agent in Charge had to go out of town and I took over the Screening Section."

Rule #3 is:

- *Ask motivator-type questions that tend to give the applicant a chance to provide revealing answers.*

For example:

Interviewer: Think of a time while at Bignam's when you really felt good . . . a time you consider a real highlight of your time there.

Applicant: I think the high point of my time there was when the section's Special Agent in Charge was obliged to return

to the Midwest on a personal leave . . . death in the family
and some estate problems . . . and I was appointed acting
supervisor during his absence. To think the Director had
that much confidence in me, well . . . I really felt good
about that.

Now the interviewer can probe that answer with the question,
"Why would the Director's expressing confidence in you make you
feel so good?" Probably the reply would be something to the effect,
"I'd worked hard and wanted more responsibility and the Director
felt I could handle it."

The original motivator-type question has revealed the following:
the applicant responds to recognition, he is an achiever, and he seeks
increased responsibility. These are very important factors to look for
in the recruiting process of the department.

Rule #4 is:

- *Ask him what* he *likes to do most on the job.* Most applicants
  do not have a chance to even consider what they would like
  to do. It is often surprising how wide a variety of talents and
  skills can surface in response to such a question.

Rule #5 is:

- *Do not waste precious time "selling" your company or department.* By the time he gets there, the applicant has convinced himself of the desirability of the job, although he may
  have a few questions he would like answered.

Rule #6 is:

- *At the conclusion of the interview, give the applicant a date
  that he can go by.*

For example, "I'd like to have you talk with my boss Friday
afternoon." Or, "Our interviewing concludes on Friday. After that
we will make our final selection. You can expect our decision no later
that next Wednesday."

Most applicants are keyed-up and nervous prior to and during
the early stages of the interview. In the security profession, which
includes interrogation responsibilities, these factors make applicants
easy prey to the experienced supervisor or manager. As the level of
professionalism in security rises, it is to be hoped that all employment interviews will be handled in a sensitive and empathetic
fashion.

Secondary Interviewing

The primary or initial interviewer in the Security Department is the person who will be the new employee's supervisor. If this supervisor has meaningful responsibility in his assignment, and if he is to conduct meaningful employment interviews, then *he* should make the selection. Why, then, should there be a secondary interview?

The secondary interview should not be a NO or GO test, but rather a consultive arrangement. The supervisor should understand that he will decide whom he wants out of all the applicants. But management must provide a climate wherein the supervisor not only wants to extend the *courtesy* of having his selected applicant meet the manager, but also sincerely wants the input, opinion and concurrence of his superior. Admittedly, this is a fine balance. But if the climate is right, the manager can actually reject an applicant with the supervisor's total concurrence and support.

Ideally, the arrangement should go something like this:

Supervisor: Out of six applicants I found a guy I really like and think he'll do the job. Before I go back to Personnel, I'd like the benefit of your thinking. I think you'll agree, but who knows, maybe you'll see something I missed. Can you talk to him?

Manager (following the secondary interview): Your candidate has very impressive credentials and I think you made a good choice.

Or

Your applicant has impressive credentials and I believe he'll be a good man. But did you realize the guy is quite inflexible in terms of transfers or promotions out of town due to his mother's health and her dependence on him? How would his inflexibility affect you?

If the information about the dependent mother strikes the supervisor as news, he might very well reconsider his selection. On the other hand, he might decide to hire the applicant anyway and modify his developmental strategy for the applicant somewhat, knowing his restrictions which probably are of relatively short duration. The point is that the purpose of the secondary interview is to confirm the interviewer's choice and/or to apprise the interviewer of additional data to consider. This is not to say that the secondary interviewer should be powerless to override the other interviewer's choice, because under some unusual circumstances such authority may very well have to be exercised.

The secondary interviewer may also be instrumental in selecting one of two good candidates. This happens when the interviewer has narrowed the field down to two and cannot decide which one he prefers.

Selection of the Best Candidate

If the interviewer understands the job function for which he is recruiting, and if he knows precisely what job qualifications are necessary (in terms of acquired skills, experience, education, and even temperament and personality), and if the interview is conducted in an objective manner, then one candidate should stand above the rest. The goal is one of objectivity; the problem is *subjectivity*. Too frequently the best candidate is not selected because of bias on the part of the interviewer. This is not the proper textbook to delve into the problems of personal bias, but it is important to observe that the professional manager must recognize that personal bias exists and does tend to distort our decisions. Overcoming such bias can be an exciting challenge.

To discipline oneself in the interviewing and selection process to look for the candidates with the best qualifications, regardless of their color, age, sex, hair style, complexion, weight, shoe size, etc., will materially contribute to the selection of the best candidate.

Background Investigation of Applicant

The subject of applicant background investigations is discussed in some depth in Chapter 16. Our purpose here is to emphasize the criticality of background screening in security. The management team that fails to turn over every possible stone in clearing a candidate for a Security Department position represents derelict management. The need to conduct a neighborhood check as part of the background investigation cannot be overstressed.

Certainly a good portion of background checking is done by telephone, but in this case the investigator should get out from behind his desk and into the field, talking to people about the candidate.

The modern security administrator knows that there is a wide range of company and personnel management reaction to employee misconduct, *except* if the employee is a member of the security staff. With any other employee, the reaction ranges from "forgiveable" to "intolerable"; with a security employee, the reaction is universally "intolerable." Negligence, gross inefficiency, misconduct or dishonesty are magnified when they are discovered within the security organization, and perhaps they should be. Since this is the case, maximum effort must be made to insure that each new member of the security staff *is* who and what he says he is.

Job Offer

Once the applicant has been chosen and screening is completed, we have come full circle back to the Personnel Department. The selected candidate's application and the interviewer's comments are reviewed by the personnel representative and interviewer. The salary and starting date are agreed upon and the matter is then left in Personnel's hands. They will make the job offer. If for any reason there is a problem with the starting date or salary, the personnel representative will serve as the intermediary until the matter is resolved. This is an important service that shields Security from what could be a disagreeable or unpleasant dialogue if handled directly between the applicant and the Security Department.

Summary

An organization is people, and the performance of the Security Department will benefit from care and attention to personnel selection.

Recruiting activity will be adapted to the job level, with open ads recommended for entry level positions. Blind advertisements, followed by more detailed screening, will be used for higher level positions.

Interviewing is the heart of personnel selection. After the initial screening by the Personnel department, the primary interview should be conducted by the supervisor for whom the selected candidate will work. Questions should require meaningful answers (not signaled in the form of the question), and should be designed to allow the candidate to reveal as much of himself as possible. A secondary interview by the Security Manager is advisable, not to overrule the supervisor but advisory in nature.

If the interviewing process is based on clear knowledge of the job function and the qualities necessary for its performance, the selection of the right candidate is often easy. Above all, that selection must be based on *objective*, not subjective, criteria.

Background screening prior to the final job offer is critical because of the responsible role played by the security officer.

Review Questions

1. What are the six steps in the hiring process?
2. Besides newspaper advertising, what is another method of soliciting applicants for entry level positions?
3. Discuss the differences in the approach to recruiting for a non-entry level position vs. an entry level position.
4. Discuss the six rules of interviewing employment applicants.
5. What are the reasons for the secondary interviewing of an applicant by the Security Manager?

Chapter 8

Training

Without question the primary contributor to poor job performance is inadequate training. Although the value and absolute necessity of sound training are extolled by all, training dollars seldom materialize. Other demands on the organization seem to push training activities down the list of priorities, and training—real, formalized training—is always going to happen "tomorrow." No other single organizational function gets as much lip service as training.

Shortcomings of Typical "Training"

Further complicating the dilemma of training is the fact that training means different things to different people; the function itself is misunderstood. The following story might serve as a typical example.

Harry X receives a phone call from a representative of the Personnel Department at Company B, where he was interviewed several days earlier. Harry X accepts the job offer. He is pleased and promptly calls his friends and relatives to tell them about his good fortune. He is to report to work on the following Monday.

When Monday morning comes, Harry X is up early. He has been anticipating this day, and he dresses to look his very best. He heads for work, sensing excitement. As the building comes into view he starts to feel some nervousness and anxiety. The tension increases as he enters the personnel offices and is greeted by a somewhat impassive personnel employee who coldly directs him to complete more forms.

Harry is then herded into a "Training Room" with a number of other new and equally nervous employees, where they meet a "Training Officer" who either acts bored or is so enthusiastic that he arouses skepticism in the minds of the new employees. Each new employee is given a company booklet which is recited to them page by page as if no one could read. The booklet describes benefits, whom to call or what to do if one is ill, retirement programs, the history of the company, and major important rules.

This concludes phase one of Harry's "training." Of course, it is not training, it is orientation. Harry and his peers learned little and will retain less, for three reasons: (1) they were not prepared for the presentation; (2) they were given too much information in too short a period of time; and (3) they cannot relate the material to their work.

Harry's state of mind at this point is becoming negative. What he is hearing goes in one ear and out the other. He wants to get to the job he was hired for. He wants to see the area he will be working in, and he wants to meet his supervisor and the people he will work with. But these personnel and "training" people (training historically falls within the personnel division) will not let him go.

Phase two of training is usually a tour of the facility. There is a welcoming address by the vice president and a luncheon in the company cafeteria. By now Harry has made friends with another new employee, and they secretly concur in their negative reaction to every company representative to whom they have been exposed so far.

Finally Harry is directed to his new department and introduced to his supervisor, who, because he was not involved in the selection of Harry, eyes him coolly and suspiciously. Harry is disappointed, hurt, disgusted in part, and close to anger. Typically, the supervisor is very busy. He does make time for the new man up to a point where he calls in one of Harry's peers and charges him with showing Harry around. Thus concludes the first day of "training."

The following day, Harry starts "on-the-job training" on the lobby desk with Frank. It is a good place to start Harry because Frank needs help; he is very busy because of the badge conversion.

Frank is annoyed because he knows an inexperienced employee will only be a hindrance. He had asked for an experienced assistant. He makes it a point to let Harry know how he feels.

Harry stands by helplessly, not knowing what to do or how to respond to questions, and attempting to avoid criticism. Midway through the confusing and distressing day there is a lull. Frank finally decides to accept Harry and begins to confide in him. He tells Harry everything he does not like about the company, about the department, about supervisors. He also passes on everything he learned (and he learned the same way Harry is now learning). He even advises Harry what company procedures (as Frank interprets them) are to be followed and which are to be ignored. The second day of "training" ends.

On the third day, Harry is assigned to the receiving docks to work with another officer because "the lobby desk is too busy."

New employee training resembles Harry's experience in far too many cases.

There are many conspicuous problems and lessons in the foregoing. Elaboration is unnecessary, with one exception: a very important and powerful lesson to be learned is that first impressions made upon the employee on his first day in a new job have great impact. The new employee is disoriented, somewhat frightened, nervous, self-conscious, and subconsciously he is crying out for a friend. The manager who recognizes and is sensitive to this can treat the newcomer in such a way as to quickly establish respect and loyalty that otherwise may not be developed.

Training Defined

As stated earlier, the training function means different things to different people; it is widely misunderstood. Certainly there is a question of definition, and a typical dictionary definition *(Webster's)* tells us little when it describes training in this manner:

"training. *Noun.* Act, process, or method of one who trains; state of being trained. *Adjective.* That trains; used in or for training; as, a *training* ship for sailors."

Even aside from its obvious circularity, what does this explanation really explain? Is it any wonder there is confusion?

A more workable definition might be the following:

Training is an educational, informative, skill development process that brings about anticipated performance through a change in comprehension and behavior.

Basically there are three things that management wants the employees to know. It is important for them to understand

1. *what* management wants them to do.
2. *why* management wants them to do it.
3. *how* management wants it done.

"POP" Formula: Policy, Objective, Procedure

Interestingly enough, the *what, why* and *how* are related to *policy, objectives,* and *procedures.* From this correlation the author has developed the POP Training Formula as the basic building block for job training.

The area of the Why/Objective in Figure 8–1 deserves special attention. Too frequently the training process overlooks the necessity of informing employees *why* this should be done, *why* that should not be done, etc. When employees are informed as to the why's, their performance will improve. This point cannot be overstressed.

What Management Wants Done.	→ POLICY	→ Education	→ Employee knows what is expected of him.
Why Management Wants It Done	→ OBJECTIVE	→ Information	→ Employee understands why he is doing his job.
How Management Wants It Done	→ PROCEDURE	→ Training	→ Employee knows how it is to be done.

Figure 8-1. The POP formula for training.

Incidentally, those who are familiar with the *who, what, where, how, why* and *when* journalistic formula may wonder what has happened to the *who, when* and *where*. The *who* (the employee who is being trained) is obviously implied, and the *when* and *where*, in this context, are included in the *how*.

Now let us translate the POP Formula into training for a specific job, such as Shoplifting Detective.

Re-examining the suggested definition, it will be seen that there

	COMPANY PROGRAM	EMPLOYEE FUNCTION
POLICY (What Management Wants Done)	Arrest and prosecute every shoplifter.	Has been hired by the company to specifically detect and apprehend shoplifters.
OBJECTIVE (Why Management Wants It done)	Reduce shoplifting losses. Deter others by example of arrests. Punish or discourage offenders.	• Helps to reduce losses caused by shoplifting. • Deters others from shoplifting. • Helps to punish offenders through the criminal justice system.
PROCEDURE (How Management Wants It Done)	Lawful gathering of the necessary evidence to justify arrest and support prosecution of shoplifters.	• Sees customer approach merchandise. • Sees customer select merchandise. • Sees secretion of merchandise. • Sees that no payment is made. • Sees removal of merchandise from store. • Approaches customer and says, "Excuse me, etc." • Carries out arrest with justification. • Makes written report of incident, etc.

Figure 8-2.   The POP formula for Shoplifting Detective training.

	COMPANY PROGRAM	EMPLOYEE FUNCTION
POLICY (What Management Wants Done)	Control entry and egress of all persons to the facility.	Ensures that no person shall enter the facility without an authorized badge. Ensures that no equipment, materials or supplies may leave the facility without authorization.
OBJECTIVE (Why Management Wants It done)	Prevent losses from theft. Prevent trespassing. Safeguard persons and company assets.	• Helps to reduce losses from theft. • Helps to prevent unauthorized access to the facility for malicious purposes (bombing, vandalism, theft of information, etc.) • Helps to make facility safer for personnel.
PROCEDURE (How Management Wants It Done)	Implementing access control program (employee and visitor badging, sign-in registers, package inspection, etc.).	• Permits entry after exhibition of authorized badge. • Refers lost or forgotten badge cases to personnel. • Prior to 6:30 a.m. and after 6:30 p.m. requires signature on registry before permitting access to facility. • Confirms visitors by telephone, issues visitor badge and awaits escort. • Inspects all containers not displaying "Security Parcel OK" slip, etc.

Figure 8-3. The POP formula for Lobby Desk Officer training.

are three aspects of training, involving education, information and skill development. The example in Figure 8-2 makes it apparent that, of the three tiers of the formula, the How or Procedure tier addresses itself to *skill development*, while the other two tiers, Policy and Objective, are *educational* and *informative* in nature. It is not enough for the detective to know that management's policy is to arrest and prosecute every shoplifter. He must also understand the objectives which make this policy reasonable and necessary, and he must thoroughly grasp the various procedures which are essential in the detective's own execution of his responsibilities. The proper combination of education, information and skill development round out and give substance and definition to "training."

Detailed Expansion of Procedures

Although each tier of the training formula is important and interrelates with the others, the Procedure tier will receive, by far, the most attention. Consider another job classification, that of Lobby Desk Officer (Figure 8-3). Management's policy of controlling access and unauthorized removal of equipment, materials or supplies, and its twin objectives of preventing trespassing and theft, are quickly grasped. What remains, for the new employee, is the question of how to carry out his assignment—how to do the How. Once on this track, the skill development process of training is well on its way.

Consider, for example, just one aspect of the procedural steps:

- *Confirm visitors by telephone, issue visitor badge and await escort.*

How should this be done? The following might be one acceptable procedure:

1. Request visitor to complete Visitor Card, form s647.
2. Use company directory to call employee whom visitor states he wishes to see.
3. If employee wishes to see visitor, request employee to come to desk to escort visitor.
4. Issue yellow Visitor Badge to visitor and record badge number on visitor's completed Visitor Card.
5. Place Visitor Card in Visitors Aboard box.
6. Invite visitor, by his name as indicated on Visitor Card (e.g., Mr. Jones), to have a seat until his escort arrives.

The detailed expansion of procedures, then, or the "How-to-Do-the-How," will be the primary thrust of the training efforts. It is, nevertheless, only a part of the whole—a large part, granted, but still a part.

Training As Ongoing Responsibility

Up to this point we have been discussing individualized and specific job description training. As critical as this type of training is, it is only part of the entire training picture. The training function within the security organization should be continuous and ongoing. Ideally, training should be under the direction of a Training Officer whose *sole* responsibility is Security Training.

The last person who should be placed in the Security Training Officer function would be a trained security officer. To insure total training objectivity, an experienced trainer or a bright college graduate with an academic background in personnel, communications or teaching should be hired and charged with the responsibility of coordinating and administering the training program in the Security Department. Such a person approaches the job without preconceived notions, without bias, without an "expert's" point of view. Rather, he goes about assignment after assignment with wide-eyed, unabashed curiosity and amazement (which tends to be contagious), learning as he goes and seeing many things the experienced security officer does not see. Such people make outstanding training officers.

Note that this opinion differs from our recommendation in Chapter 3 on the conduct of training sessions on security for the *general* employee. In such sessions only the security employee can speak with authority about the security function.

Types of Security Training Programs

Following is a list of types of training programs which security management could provide the department's employees:
- *General Seminars*
  Usually most effective if by employee classification; e.g., all patrol and uniformed personnel, or all fraud investigators. So-called "general" programs are a potpourri of subjects that are important and meaningful to the group. One side effect

of these sessions, which could be from one to three days in duration, is the motivational aspect, which should be capitalized upon in the agenda. Following is a typical agenda:

9:00– 9:30	Welcome and Introduction	Security Director
9:30–10:15	Organizational Overview (transparencies and handouts)	Security Manager
10:15–10:30	Break	
10:30–12:30	Inter-personal Communications	Manager, Public Relations (from outside the department)
12:30– 1:15	Lunch	
1:15– 2:00	Do's and Don't's in Handling Company Employees	Training Officer
2:00– 3:30	Report Writing (work sheets and handouts)	Chief Investigator
3:30– 3:45	Break	
3:45– 4:30	Training Manual Update (revised pages)	Security Manager
4:30– 5:00	Open Question-and-Answer Period	Staff
5:00– 5:45	Film, "You Pack Your Own Chute" (16mm projector)	Training Officer

This outline illustrates how flexible a general seminar can be, depending upon the objective. In this agenda, for example, the real or *primary* objective of the session might have been to introduce and explain a major organizational change for the Security Department. This objective was achieved in the morning session. Because people had to come from far and near, the balance of the day was then devoted to training.

• *Interrogation Workshop*
  Half- to full-day session of principles and techniques of interrogation with role-playing and, ideally, video play-back of role-playing.
• *Testifying in Court Seminar*
  Half- to full-day program which includes preparation of evidence, dress, demeanor on the witness stand, voice, where to look, "traps," stress and attitude.

- *Report Writing Workshop*
  90-minute review up to a full day on the principles to be
  followed in reducing events to paper.
- *Supervisory Training*
  Two-hour "topic" sessions up to full three-day seminar.
  "Topic" session could be *How to Handle Disciplinary Problems,* or *Management Styles, X, Y or ?.*

Meeting Organizational Needs

The types of training programs are limited only by organizational needs. True, much material is available at local universities
and community colleges, and security personnel should be encouraged to further their education at such institutions. Specific organizational needs, however, usually must be met through "in-house"
education.

Organizational needs come down to people needs. The agenda
of the general seminar discussed above was in response to needs.
Security management had a need to communicate the reasons for
and details of a major reorganization. Security management also
recognized the daily operating problems connected with poor inter-
personal communication skills between security people themselves
and between non-security people and security. Therefore a second
need was addressed in the program. The remainder of the agenda
was the result of a survey (made in advance by the Training Officer)
of the employees' stated needs. Thus, training objectives are identified and material designed to achieve those objectives. Too many
training programs are the master works of art of training personnel
or management, but they miss the mark of satisfying what employees
want and need.

Security Training Manual

A Security Training Manual or Handbook is an absolute essential, and it must be updated on a regular basis. The subject matter
should include pertinent company policies, departmental policies,
job descriptions, emergency phone numbers, and a great many procedural instructions for specific incidents, such as telephone bomb
threat or facility black-out. In some organizations the manual is
deemed sacred and consequently most employees are not allowed to

touch it—a foolish attitude. The manual should be put in the hands of all regular security personnel.

Summary

Typical new employee "training" is little more than general orientation. The newcomer is usually unprepared for the presentation, and he is given too much too fast, in a manner unrelated to the work he will actually be doing. The typical orientation program, in which the employee is lumped with many others from various departments for a day and then thrown immediately into an assignment for "on-the-job" training, is actually negative training. The employee learns the wrong way—or the wrong things—and develops undesirable attitudes.

Effective training is an *educational, informative, skill-development* process. The basic building block for training can be summed up in the POP Formula:

Policy   = *What* management wants
Objective = *Why* management wants it
Procedure = *How* management wants it accomplished

Although the importance of why something must be done cannot be overstressed, primary attention in training will be on the procedural, or skill-development, phase . . . learning "how to do the how."

Training should be an ongoing process, ideally in the hands of a Training Officer selected for qualities other than security experience. Types of training programs may include *general* seminars and *specific* seminars. They should be based on specifically identified organizational needs.

The Security Manual should embody the essentials of security responsibilities, and should be in the hands of every employee.

Review Questions

1. What is a workable definition of *training?*
2. What are the three basic things that management wants employees to know? How does the "POP Formula" relate to these?
3. List four possible topics for security training seminars.
4. What should the contents of a Security Training Manual include?

Chapter 9

Discipline

As a rule, the very word *discipline* evokes an emotional reaction on the part of employees at all levels of the organizational pyramid. Most supervisors and managers would rather do anything but discipline, and it is human nature to resist and resent punishment. This negativism surrounding a critically important organizational process is unnecessary and can easily be replaced with a positive approach, called Constructive Discipline.

Before discussing Constructive Discipline, it is instructive to consider a number of dictionary definitions of the word *discipline*.

1. Training that corrects, molds, or perfects.
2. Punishment.
3. Control gained by obedience or training.
4. Orderly conduct.
5. A system of rules governing conduct or practice.
6. To punish or penalize for the sake of discipline.
7. To train or develop by instruction and exercise.
8. To bring a group under control.
9. To impose order upon.

The majority of these explanations emphasize punishment or control, both of which are aspects of discipline. Only the first and seventh in this list call attention to the key aspect of Constructive Discipline: *the training which develops disciplined conduct.*

The word *discipline* is derived from the Latin *discipulus* ("learning"). The word *disciple* also comes from the same root; the early Christian disciples were considered "students" of Christ. The origin of the word suggests this important concept: that positive and constructive discipline is *training that corrects, molds, or strengthens* an employee in the interests of achieving departmental and company goals. Punishment, that factor which is feared and disliked by all, is secondary. Any punishment connected with discipline should always be a *means to an end,* and that end should be organizational, not personal.

In other words, the effective disciplinary process condemns the act, not the employee. It says, "You're okay, but what you did is not okay." By addressing ourselves to, and focusing on, action rather than personalities, the whole process takes on a constructive dimension that is easy to handle and acceptable to all. Comprehension and subsequent application of this positive concept have helped many managers cope with their disciplinary problems.

It is also important that discipline be swift. The long-range effect of coming to grips with a problem immediately is better than putting off what probably will have to be faced later—irrespective of the nature of the problem, be it simple tardiness or a careless oversight. What could easily be corrected now may be far more difficult to correct later, because the real essence and secret of constructive discipline is its *preventive* nature. To train, mold, and correct *now* reduces the need for more difficult training, molding, and correction later.

The Supervisor's Role in Discipline

Discipline is a responsibility which rests squarely on the supervisor's shoulders. It cannot be passed on to a higher supervisor and should not be passed on to the Personnel Division. Some weak supervisors shirk their responsibility with the idea that enforcing the regulations will hurt their relations with subordinates. Actually, most people prefer to work in a well-ordered environment. They really do not expect or necessarily want the supervisor to be too lenient all the time, because those who fail to exercise needed discipline, who will not say "no" or "don't" to those who deserve it, can make the workload more difficult for everyone else.

The supervisor who is fair and consistent in his treatment of employees will gain rather than lose respect through being firm and expecting conformity to the rules. Once the proper atmosphere is created through constructive discipline, a request from the supervisor *is* an order. There is no need to be abrupt or overly forceful to get the job done, because employees will respect the supervisor who respects them.

Some make the mistake of believing that discipline is only directed at the inefficient worker. *All* employees require constructive discipline. There are times when disciplinary action is essential with an outstanding employee, usually because he and others have come to think he is so good that he is indispensable. The supervisor should never lose respect or control of the organization by being afraid to lose a good man.

The supervisor who understands the employees' psychological needs will generate less reactive hostility, and consequently experience less resistance, than the supervisor who approaches the employee with insensitivity and harsh tactics. An important key is to recognize the individual differences of employees, handle them on that basis to win their loyalty and support, then motivate them to greater personal success. The benefit will be a significant reduction in disciplinary problems.

All disciplinary actions commence with an interview and discussion. If handled with sensitivity (which includes understanding the employee's psychological needs and treating him or her as an important individual), the interview can accomplish its basic purpose and at the same time actually serve to improve the personal relationship between the employee and supervisor. The employee frequently expects the worst. With many supervisors the employee would leave the interview feeling misunderstood, mistreated, hostile, guilty, or dejected. If the supervisor remembers that the basic purpose of discipline is correction and training, not punishment, he will take a positive approach in the interview. That approach will leave the employee with renewed confidence in himself and in the supervisor and greater faith in and respect for the boss's good judgment and fairness.

Disciplinary Problems Arising
From Misunderstood Assignments

The following sign hangs in a number of supervisors' and managers' offices:

I KNOW YOU BELIEVE YOU UNDERSTOOD WHAT YOU
THINK I SAID, BUT I AM NOT SURE YOU REALIZE THAT
WHAT YOU HEARD IS NOT WHAT I MEANT.

The irony here is that the statement is an absolute indictment against the supervisor. When the subordinate has failed to do a task as assigned, and his superior proudly directs him to read the sign, the supervisor obviously fails to recognize his own responsibility to make each assignment clearly understood. The failure really rests with the supervisor, not the subordinate. Many disciplinary cases are the result of assignment failures. And most assignment failures have nothing to do with the employee's level of competence but rather with a misunderstanding of what was expected.

How many times has a supervisor stopped an employee, given instructions on what he wanted done, and then asked, "Do you understand?" The employee nods knowingly, but as soon as the supervisor walks away, the employee turns to a peer and asks, "Do you know what he wants?" The supervisor's first error is in asking the employee if he understands. Most employees will say yes rather than admit they failed to grasp the instructions.

Other assignment errors include:

- Instructions may not have been given in a logical order or sequence.
- The person giving the instructions may have spoken indistinctly or failed to use clear language.
- Instructions may have been too complicated for one simple explanation.

There are occasions when an assignment is indeed understood and yet still not followed because of the manner in which the assignment was given. The ideal way to give an assignment is by a request rather than a demand. Asking an employee to do something makes him part of the picture and gives him more opportunity to make suggestions and to feel a responsibility to perform the assignment. Requests create a spirit of willingness to do the job.

Following are ten suggestions to follow in giving assignments:

1. Know the assignment yourself.
2. Do not assign work above the employee's ability.
3. Explain the purpose of the assignment so that the employee understands why he is being asked to do it.
4. Request or suggest—do not demand. For example,
   "Would it be possible?"
   "Suppose we try it this way?"
   "Will you take care of . . . ?"
5. Give brief, exact instructions with all of the necessary details, but not too much to confuse.
6. Demonstrate if possible.
7. Do not assume the employee understands. Have him reiterate the instructions.
8. Do not watch every move; let the employee feel responsible.
9. Let the employee know you are there if he needs assistance.
10. Be certain these points have been covered:
    a) Who is to do it.
    b) What is to be done.
    c) Where it is to be done.
    d) When it is to be started and finished.
    e) How it is to be done.
    f) Why it is to be done.

Most employees want to do a good job. If care is taken in giving assignments, there will be fewer failures and fewer disciplinary problems resulting from failures.

Basic Rules of the Disciplinary Process

There are six fundamental rules in the disciplinary process that have universal applicability.

Rule #1—*Put rules in writing* and make certain employees understand them. There should be no assumed rules. If a rule is worth having, it is worth writing down. Employees are entitled to know what the rules are if compliance is expected.

Many institutional rules are peculiar to the organization and therefore not of common knowledge, particularly to someone new to the organization. Take, for example, the situation of a security officer who forgot his badge and, being new to the business, borrows a badge from another new officer who is going off-duty. Experienced personnel would appreciate the logic behind prohibiting an officer from wearing another officer's badge, but a newcomer might not understand. If he is to be held accountable, he should know the ground rules. Many firms provide new employees with a copy of the rules and then have them sign a statement to that effect which becomes part of their personnel file. Other companies post rules in conspicuous areas such as employee locker rooms.

Not only is it morally wrong to take punitive action against an employee who was honestly ignorant of a given rule, it is an administrative or legal wrong which can be remedied by the government, especially if the punitive action is termination. The company may be legally bound to reinstate a wrongfully discharged employee, *with full wages for all time lost due to the discharge.* Many companies have paid months of back wages under such circumstances.

In short, there must be no surprises in terms of company rules.

Rule #2—*Discipline in the privacy of an office.* To the employee, being corrected for deficiencies in conduct or performance is a sensitive and frequently embarrassing experience. To be corrected in the presence of others is considered degrading, and the end result of that approach is seething resentment and angry embarrassment— emotions which are counterproductive to the true disciplinary goal. Additionally, the privacy afforded in an enclosed office permits the participants to hear each other clearly. It is extremely important for the supervisor to hear what the employee has to say and for the employee to hear what the supervisor is saying.

Rule #3—*Be objective and consistent.* As stated at the outset of this chapter, effective discipline condemns the act, not the person. That approach is obviously objective; the issues are not, or at least should not be, personalities. The supervisor who refers to an employee as "dummy," or who makes such statements as "Can't you get it through your thick skull . . .," or who succumbs to personal likes and dislikes, loses objectivity and consequently loses credibility and respect. Thus the supervisor is no longer practicing true Constructive Discipline but returns to the negative approach of punishing people who fail to meet standards.

Inconsistency is equally deadly. If the policy of the department is to terminate officers who sleep on the job, then all officers so caught must be terminated. To fire one and not another breeds contempt for the management of the organization. Conversely, if the same rule is consistently enforced and acted upon, genuine respect, not only for the rules, but for the management, follows.

Rule #4—*Educate, do not humiliate.* The concept here is to *help, not hurt* an employee who has failed to meet standards of conduct or performance. If the disciplinary action truly corrects, trains or molds the individual to meet standards, he comes away from the experience with better insight into himself and what the company expects. He comes away educated. If he is berated and humiliated, he comes away angry and resentful and certainly destined to fail again, sometimes by design. Both he and the department suffer as a consequence.

Rule #5—*Keep a file on all employee infractions.* This is not to suggest that a negative dossier be maintained on each employee. Rather, documented incidents of past failures are a necessary and useful reference for repeated incidents. Compare, for example, the two following situations.

#1) *Supervisor:* John, I've talked to you before about your uniform, and you're out of uniform again with those blue socks.

John: You've never mentioned my socks before . . . when was that?

(Supervisor recalls the incident only vaguely, and has no documentation of it.)

#2) *Supervisor:* John, I've talked to you before about your uniform, and you're out of uniform again with those blue socks. In fact, last February 15th we talked about blue socks, and again on March 30th I talked to you about your shirt cuffs being turned up. This being the third time, I'm going to place a formalized written reprimand in your personnel jacket. You must understand that we insist on every officer dressing according to the uniform code. A uniform improperly worn lacks good taste and is a poor reflection on the organization as a whole.

(The facts are obviously clear and available, and John cannot challenge those facts.)

The "file" this rule refers to is an informal record, maintained by the immediate supervisor for his own personal reference. Dates of incidents may or may not end up as formal documentation in the personnel records, depending on the employee.

Not only does the employee find it difficult to argue with these supervisorial records, so does any administrative hearing board which may someday sit in judgment over the company's more drastic disciplinary action against the employee.

Rule #6—*Exercise discipline promptly.* Consider again the situation of John and his uniform violation of wearing blue socks when he is supposed to wear black. This time, the supervisor has delayed his corrective action.

Supervisor: John, I've talked to you before about your uniform, and last week you were out of uniform again by wearing blue socks.

John: I don't recall wearing blue socks last week. What day was that?

Supervisor: Well, *I* recall your wearing blue socks, and it happened to be last Friday.

John: If I was wearing blue socks they were navy blue and no one would ever know the difference.

Supervisor: They weren't navy blue, they were bright blue— obviously and conspicuously bright blue.

John: Did you see them yourself?

Supervisor: I did.

John: Well, then, why didn't you talk to me about them *then* if they were so bad?

Good question, John!

It should be apparent that with the passage of time (and distance) from the infraction, whatever magnitude it may have, the issue becomes vague and almost argumentative. If corrective action is appropriate, then it must be handled *now* or on as timely a basis as possible. It is like catching a child with his hand in the cookie jar. If prompt action follows the detection, the youngster can relate the consequences of his action to the act itself. If he must wait until his father gets home after work, it is harder for the child to make sense out of his scolding or punishment. Besides, depending upon the level of sophistication of the youngster (or adult), he will rationalize the incident to minimize its importance, and subsequent corrective action can appear unreasonable.

The same concept of the need for prompt action is most evident in dealing with such drastic behavior as theft. If you observe an employee steal and immediately take that employee into custody and commence the interrogation, the interrogator has all the advantages and will experience little in the way of resistance. If that employee is allowed to leave the site, however, and is not interrogated until the following morning, resistance and obstacles will surface. Yet the same act occurred. The difference is timing. Delays raise questions of credibility.

Self-Discipline

No manager or supervisor can ever hope to discipline others effectively if he cannot discipline himself. Disciplining oneself can be accomplished by controlling vanity, likes and dislikes, personal thoughts, and by always being humble. Self-discipline will lay a solid groundwork for working with other people and their failures and problems, and for setting a climate where self-discipline becomes contagious.

Self-Discipline and Vanity. The supervisor who misuses authority will earn resentment instead of respect. Barking out commands may seem the quickest way to get the job done, but that technique is a vain self-indulgence that a manager of people cannot afford. Using power in this way is not leadership. Everyone knows the extent of the manager's power; it need not be displayed. Self-control over such vanity is also detected by subordinates, and in their own behavior they will respond in kind to that example of personal discipline.

Self-Discipline and Temper. Loss of temper may make a manager feel better for a while, but it will not improve his performance in handling people or assignments. People never know how much to discount instructions issued in an intemperate manner.

Self-Discipline and Arguments. Most arguments are useless. Discussions, not arguments, produce agreement and cooperation.

Self-Discipline and Personal Likes and Dislikes. Nothing creates a better atmosphere than friendly recognition of subordinates on an equal basis. And nothing creates trouble faster than failures to control personal likes and dislikes and developing personal favorites (including outstanding employees) or exhibiting personal prejudices and dislikes. Real self-discipline is required to deal with all personnel with honest objectivity.

Self-Discipline and Work Habits. Subordinates cannot be expected to discipline themselves in terms of good work habits if the example set by management is one of poor work habits. The manager must discipline himself to be punctual, timely with his assignments, thorough, orderly and accurate, knowing that subordinates notice far more than one might suspect. Supervisors live in glass houses on the job.

Self-Discipline and Humility. The effective manager will never hesitate to acknowledge his own errors. He is not going to be right every time, and the rest of the organization knows it. He should not be embarrassed to say, "I don't know." Although it requires self-discipline, the manager should not be hesitant to ask others, including subordinates, for their opinions, knowing that they may have some ideas better than his own.

Summary

Discipline is training that corrects, molds and strengthens an employee, at all levels in the organization, in the interest of achieving departmental and company goals. Constructive discipline is *positive,* focusing on correct action rather than on personalities; on the act, not the employee.

In the security organization, discipline is primarily the responsibility of the supervisor. Effective discipline begins with effective communication and full understanding of what is required. Its purpose is *corrective training,* not punishment.

The basic rules of the disciplinary process are: (1) put rules in writing; (2) discipline in privacy; (3) be objective and consistent; (4) do not humiliate the employee; (5) keep a record of infractions and disciplinary action; and (6) exercise discipline promptly.

Effective discipline will find its model in the Security Manager's own self-discipline and restraint.

Review Questions

1. Define *constructive discipline*.
2. Give several possible reasons for misunderstood assignments.
3. Discuss the six basic rules of the disciplinary process.
4. Why is it a good idea to keep a file of employee infractions?

Chapter 10

Motivation and Morale

The question of how to motivate employees to do more and better work and how to keep them happy and interested in their work remains a constant challenge to management, a project for researchers, and a thesis subject for academicians. What motivates one person may not motivate another. Motivators that are effective in one industry may be out of the question in another. For example, you cannot motivate a security officer to write more citations in order to win a free trip to Hawaii, but that type of motivation is used in real estate and other types of sales companies.

There is even disagreement over whether that free trip to Hawaii is a motivator. Some management theorists argue that it is an inducement or a carrot on the end of a stick that keeps employees producing at levels predetermined by management. Others say, "Call it what you will, it gets the job done, and when he gets to Hawaii, his morale will be way up there." An opposing view holds that, if you take away the "carrot," fundamentally good people will produce at the same level and poor performers will still be poor performers (they never won the Hawaiian trip anyway). The latter position suggests that perhaps motivation is internalized as opposed to externalized; that is, motivation comes from within the person and not from without. If that is the case, is it really possible to motivate another?

"Theory X" and "Theory Y"

Before we attempt to deal with that question and others about motivation, morale, and human behavior on the job, we should have some understanding and insight into classical studies that have gone before us. For example, Douglas McGregor's "Theory X" and "Theory Y" are remarkably pervasive assumptions about human nature and behavior. The first three, which still have far too much acceptance in our society today, McGregor calls "Theory X" assumptions:*

"Theory X" Assumptions

1. The average human being has an inherent dislike of work and will avoid it if he can.
2. Because of their dislike of work, most people must be coerced, controlled, directed and threatened with punishment to get them to put forth adequate effort toward the attainment of organizational objectives.
3. The average human being prefers to be directed, wishes to avoid responsibility, has relatively little ambition, and wants security above all.

In contrast to the autocratic approach to employees reflected in "Theory X," McGregor's "Theory Y" encourages managers to be supportive of their employees.

"Theory Y" Assumptions

1. The average human being does not inherently dislike work. The expenditure of physical and mental effort in work is as natural as play or rest.
2. External control and the threat of punishment are not the only means of bringing about effective organizational effort. Man will exercise self-direction and self-control in seeking to obtain goals to which he has committed himself.
3. Part of the rewards of achievement are found in the ego satisfaction and self-fulfilling aspects of the individual commitment.
4. The average individual, under the proper conditions, learns not only to accept but to seek responsibility.

*Davis, Keith, *Human Relations at Work*. Third Edition. (McGraw Hill, 1967.)

5. The capacity to exercise a relatively high degree of imagination, ingenuity and creativity in seeking to solve an organizational problem is quite widely distributed throughout the population.
6. Under the conditions that exist in today's industrial and economic life, the intellectual potential of the average person is only partially tapped.

Again, "Theory X" is *not* an unpopular theory today. Those who put credence in these three assumptions would have very little interest, if any, in motivation.

On the other hand, the assumptions of "Theory Y," certainly a positive and enlightened approach, suggest areas that might indeed "motivate" employees. But before we examine those suggested areas, let us look at another classical work, dealing this time with *organizational* behavior instead of human behavior. The three theories of organizational behavior are The Autocratic Theory, The Custodial Theory, and The Supportive Theory.*

The Autocratic Theory

The Autocratic Theory has its roots deep in history, dating back to the Industrial Revolution of the mid-eighteenth century. The theory is based on absolute power. It tends to be threatening, relying on *negative* motivation backed by power. The managerial posture is one of formal and official authority.

In practice this theory means that management knows best, and it is the employees' obligation to follow orders without question. ("Yours is not to question why. Yours is but to do or die.") Employees need to be persuaded and prodded into performance, not led. Management does the thinking and employees do what they are told. Management has absolute control over the employee. The Autocratic approach is a useful way to get work done and therefore has some merit. Though just a step above the slave-master work relationship, it was the dominant and prevailing theory until very recent times. The Autocratic approach did build transcontinental railroads, run giant steel mills, and in general produced the dynamic industrial economy of the early part of this century.

Ibid.

The Autocratic approach gets results, but only moderate results, and does so at high human costs. And the theory does nothing to develop human potential in an organization.

The Custodial Theory

The Custodial Theory depends upon company wealth to provide economic benefits for the employee. These come in the form of pensions, insurance, medical benefits, salary increases, etc. The managerial posture or orientation is towards tangible benefits. The employee relies on the company for security instead of on the boss, as in the Autocratic Theory. The aim is to make the employee happy, contented, and adjusted to the work environment.

This approach does not motivate employees to produce anywhere near their capacity, nor are they motivated to develop their full capabilities. Consequently, employees fail to feel genuinely fulfilled or challenged on the job. Thus they must look elsewhere, such as to the bowling team or any other outside activity that holds their interest.

The Supportive Theory

The Supportive Theory depends upon leadership. Through such leadership management provides a climate in which an employee may grow and achieve those things of which he is capable— to his own benefit as well as the company's. When management creates this type of supportive work climate, employees will take on responsibility, strive to contribute to the organization, and work at improving their own skills and performance. The employees in this climate tend to think in terms of "we" rather than "they," and organizational objectives become "our" objectives. Ideally, under this approach, the employee needs little supervision. The primary need is for the employee to tell his supervisor what kind of support he needs from that supervisor in order to do a better job.

Work Motivation Theory

McGregor's "Theory Y" assumptions and the Supportive Theory of Organizational Behavior are the basis of an enlightened approach to motivation. With those as a backdrop, let us now consider Frederick Herzberg's Work Motivation Theory, a meaningful and outstanding work of recent times.*

Dr. Herzberg's position is essentially that motivation—genuine work motivation—comes from the work itself, not from those factors or "things" such as salary and job security that surround the work itself. He breaks down the job into two basic categories: (1) the job surroundings, or "hygiene" factors, and (2) the job itself and its motivators.

THE JOB SURROUNDINGS	THE JOB ITSELF
Hygiene or "maintenance" factors	Motivators
• Pay	• Responsibility
• Status	• Achievement
• Policy and administration	• Recognition
• Interpersonal relationships	• Advancement
• Benefits	• Growth
• Supervision	
• Working conditions	
• Job security	

According to Dr. Herzberg, hygiene factors do not lead to work satisfaction or happiness; rather, they lead to *dissatisfaction* more often than not, whereas the work motivators are the primary cause of *satisfaction*. Further, hygiene factors are *expected;* that is to say, good pay and periodic increases are expected, and so are benefits, good supervision and good working conditions. All are *expected once they have been given.* Certainly if there had been no dental plan, for example, and the company introduced a dental plan, everyone would be pleased. Included in their conversations would be remarks like "It's about time. So-and-so's had a dental plan for years now." Next year they will expect some new benefit—perhaps eye care. The point is, employees are not motivated to produce more for any extended period of time, nor do they find satisfaction in

*Herzberg, Frederick, "One More Time: How Do You Motivate Employees?" *Harvard Business Review*, Jan.–Feb. 1968.

their daily work, as a result of factors which surround the work.

Real motivation, then, comes from the work itself and those motivating factors that are intrinsic to the job. Let us analyze how such factors fit into the security environment.

Responsibility As A Motivator in Security

Genuine responsibility is perhaps the most important motivator, and in the security environment, unlike many other types of work environments, real responsibility can be a significant factor in an employee's work. Responsibility in this context includes such things as problem-solving, decision-making and accountability.

Why is security "unlike many other types of work environments"? The answer is that relatively few departments or jobs within departments deal with unusual, erratic or criminal human behavior, few deal with accidents or other emergency conditions, and fewer still *plan* to deal with such conditions. It stands to reason, then, that the security employee who is indeed given the opportunity to solve problems revolving around behavior or emergency situations, *before or after the fact,* really has responsibility. Those security employees who are permitted or are obliged to make decisions as to what to do or not to do when criminal or unusual human behavior occurs or when catastrophe strikes, really have responsibility.

If such responsibility is built into the security job, then, one might assume that all security people, by virtue of their chosen vocation, are motivated. Unfortunately, this is not the case, because many security people are given a "sense" of responsibility, which is quite different from *real* responsibility. When the chips are down, the real decisions come from above; the problems are solved by someone higher up, and no real accountability exists. For example, an officer is placed on a surveillance position with binoculars and is told, "Harry, we have information that a thief is going to penetrate our facility sometime tonight. You are responsible for this fence line and this side of the facility. We've got to catch this guy tonight, and I'm counting on you."

Does Harry have real responsibility in this assignment? At this point it sounds as if he does, but let the supervisor complete his instructions to Harry: "If you spot the guy climbing over, cutting through, or slipping under your section of the line, immediately contact Sgt. Green on your radio and he'll get there quick and take over."

Harry has no real responsibility here; Sgt. Green has. If you want to *motivate* Harry, then give him the responsibility to capture the thief. Give him a back-up man, a peer who will respond to the radio call that penetration is being made, or match him up with a partner and charge the two with the responsibility of apprehending anyone who penetrates their assigned area. That is responsibility. They will decide when and how to move in on the thief. That problem is theirs and theirs only, and if they apprehend or lose the thief, they are totally accountable.

Similarly, giving a security supervisor the responsibility to plan for a specific segment of an upcoming special event (say the traffic flow and parking connected with the visit of foreign royalty) is, in and of itself, an exciting challenge and a motivator. But if the supervisor's manager changes the plans, not because they are wrong but because the manager prefers his way over the subordinate's (and this is common), the so-called responsibility turns out to be only a facade. In fact, it becomes a *demotivator,* inspiring indifference and breeding suspicion of future assignments.

Most security personnel crave responsibility and have the capacity to assume more than most administrators are willing to believe. Give them as much responsibility as possible and let their work motivate them to peaks of achievement.

Achievement As A Motivator

Group and collective success is certainly important to the employee who is a member of the group. But the employee, from a personal work motivation point of view, must have the opportunity to be singularly successful, even if it is but a small success or achievement. The investigator who "breaks" a case, or the store detective who catches a shoplifter, experiences the full joy of achievement. *He* did it! And each achievement tends to drive the employee on to another.

More frequently than not, achievement comes through one form or another of problem-solving. Thus, the supervisor or administrator who recognizes achievement as a motivator will provide subordinates with the opportunity to solve a problem, to come up with the answer or to catch the thief, for the motivational benefits that success brings. Opportunities, as such, are almost limitless. As an example, ask an employee if he could design a better case history form for arrestees, or an improved filing and index system for security records—any assignment within the limits of the employee's capabilities that will provide a chance for him to truly accomplish something, to be able to say proudly, "I did that," or "That's mine!"

Recognition of Achievement As A Motivator

Rare is the person who is not motivated by praise, flattery, or any other complimentary form of recognition. To say "well done" goes a long way. Not to say "well done," when it is due, goes a long way, too, but the wrong way. It is a demotivator.

Growth As A Motivator

Growth is the consequence of expanding one's horizons, increasing insight brought about by an ever-widening variety of experiences, gathering in new ideas, concepts and information, coping with new situations and problems. All of these factors increase the individual's personal and professional growth.

Thus it behooves security management to provide a work climate that not only allows for growth, but encourages it. If there is a local security seminar, send as many people as possible. If there are local security associations, encourage appropriate members of the department to join and be active. As new assignments and problems surface, do not always call on the same already proven members of the staff to handle them. Assign someone who has not had that kind of experience so he will have the benefit of the new challenge and subsequent growth. Rotate your people around the organization rather than developing specialists in narrow areas. Rotation provides for growth. Encourage non-college graduates to return to school on a part-time basis.

Care for, water and feed your people as a gardener tends his garden, and you can actually see them "grow" before your eyes. And that growth motivates them to be achievers, ever seeking more responsibility.

Advancement As A Motivator

Opportunities to move up in the department (or in another department of the company) must be clearly visible and, in the eyes of the individual, personally attainable. If vertical movement (real or imaginary) is not apparent, then the persons who seek responsibility —the achievers, the ones who have grown and are growing—will be moved by their inner motivation to seek advancement elsewhere. Ideally, advancements should be occurring throughout the organization frequently, from interdepartmental promotions to advancements to supervisory posts in other departments, to managerial positions in security departments of corporate sister companies. Such movement is motivational in and of itself.

More production, more creative contributions to the organization, more loyalty, and more dedication to excellence in performance have a better chance of actually happening in that work environment which embraces McGregor's "Theory Y," the Supportive Theory of Organizational Development, and Herzberg's Theory of Work Motivation. The department and the organization can only profit when they recognize the value of human dignity and the creative and productive potential of their employees, and then give them room to work and to breathe.

"Demotivators"

Naturally there is a very close relationship between motivation and morale. Highly motivated people enjoy good morale, and vice versa. Dr. Mortimer R. Feinberg, Ph.D., Professor of Psychology at the Baruch School of Business and Public Administration, has identified a number of factors that can have dramatic negative impact on employees. He calls these factors the Ten Deadly Demotivators. Every security supervisor, manager and administrator should be familiar with them.*

1. *Never belittle a subordinate.*

Do not subject any employee to the ultimate put-down by call-

*Feinberg, Mortimer R., *Effective Psychology for Managers* (Englewood Cliffs, N.J.: Prentice-Hall, Inc., 1966), pp. 127–131.

ing him stupid. He might apply the term to himself for some careless mistake, but the manager should not. Generally speaking, you
can call an employee almost anything else—lazy, sloppy, slipshod—
and he will accept the criticism well enough. But calling him stupid
will not only deflate his ego but will undermine his initiative. How
can a stupid man be ambitious or enthusiastic?

   2. *Never criticize a subordinate in front of others.*

   A reprimand delivered in the privacy of your office will be accepted. The same criticism delivered in front of the employee's co-
workers will breed quick resentment. You have mortified the
employee in front of his or her friends—and that is unforgiveable.
Like most of Feinberg's Ten Deadly Demotivators, this rule is or
should be familiar to all managers. When it is forgotten in a moment
of anger or haste, the damage can be permanent.

   3. *Never fail to give your subordinates your full attention, at
   least occasionally.*

   On the positive side, making each employee feel that you care
about him personally is a strong motivator. Have each employee in
your office occasionally—and give him your undivided attention.
Have your secretary hold your calls. If you allow interruptions, lose
the train of thought and have to say, "Now then, Fred, what is it
you were saying about your family?" you will be telling Fred that
you do not really care.

   4. *Never give your subordinates the impression that you are
   primarily concerned with your own interests.*

   While your personal goals may very well be your primary concern, it is a mistake to allow your employees to think that you are
"using" them for your own selfish ends. For example, if you have
a subordinate work late to finish a project that will make you look
good to *your* superiors, make a point of sharing the credit for getting
the job done. You will still look good—not only to higher management but also to your subordinate.

   5. *Never play favorites.*

   This is another cardinal rule of supervision, but—human nature
being what it is—it is one of the most frequently broken. The
moment you start playing favorites, especially when the person in
question has been playing up to you to gain favored status, you will
antagonize the rest of the staff. Dr. Feinberg cites an example that
illustrates this point well.

A supervisor regarded as "autocratic" was accused of playing favorites. He would not let any of the secretaries in his office get away with anything, except for one. He never chastised or criticized this particular secretary, even though she was notorious for spending a lot of time making personal telephone calls. His reason was that she often stayed at the office late in order to get work done. (Naturally, this was because she wasted so much time during the day.) The rest of the secretarial staff knew that the supervisor was being manipulated, and they resented it. Consequently, the girls became demotivated.*

6. *Never fail to help your subordinates grow—when they are deserving.*

When employees feel that a supervisor is on their side, and will even fight for them if necessary—in the matter of obtaining raises that are deserved, for example—they will be more loyal and more strongly motivated. Support your employees in their attempts to grow, even if it means that you might lose a good man to another department. If you stand in his way, you will probably lose him anyway. Once an employee believes you are not on his side, he will be demotivated.

7. *Never be insensitive to small things.*

Avoid loose or rash statements—you may regret them later. In one company a department manager, on being told that one worker was unhappy and might quit, responded, "Let him—we won't miss a beat"—the statement being accompanied by a contemptuous snap of the fingers. The comment was repeated and became known to the other employees, who began using it as a mocking refrain. The statement said quite clearly, "The employee doesn't matter." In another example given by Dr. Feinberg, a supervisor known for his terrible temper roared to an employee one day, "I don't care how long you've been with this firm. Seniority means nothing in my department." This particular company had been non-union for the 75 years of its existence, but, exploiting this rash comment, the union finally won a foothold. Its slogan was, "Seniority means nothing."

By the same token, consideration in small things, from remembering to inquire about an employee's wife's illness to congratulating him on his son's graduation, will increase his loyalty and desire to work for you.

*Ibid., p. 129.

8. *Never "show up" employees.*

This rule is closely related to the first two "demotivators." Just as you should not humiliate an employee by calling him stupid or criticizing him in front of others, do not show off at his expense by demonstrating how you can do a particular job better and faster than he can. The manager should be able to perform many tasks better than his subordinates—that is why he is a manager. But it is important to any employee to take some pride in his work. If you take away that pride and self-respect, you will discourage and demotivate him. Training him to do a better job is one thing; embarrassing him is another.

9. *Never lower your personal standards.*

Care and consideration of employees should not extend to the point of accepting or tolerating inept performance. This will only demotivate the real achievers in the organization. The reasons are summed up in an analysis Dr. Feinberg quotes from the Research Institute of America:

> The mediocrity of colleagues can muzzle the initiative of the dynamic doer who has high standards for his own performance . . . especially when the "mediocres" are permitted to stand on the sidelines and throw darts at new ideas. Management often tolerates a certain percentage of people whom they have given up on . . . men who will never pull their own weight. But if these people are permitted to remain in key positions, just the simple fact of their presence can cost the company the loss of an endless chain of worthwhile people who don't have to work against such odds. And, incidentally, whether they remain on the payroll or leave for greener fields, you've lost a man if he has decided it doesn't pay to knock himself out.*

10. *Never vacillate in making a decision.*

Effective management is characterized by the willingness and ability to make prompt, wise decisions. Every decision involves an element of risk—the chance that you could be making a mistake. If your employees see that you lack confidence in your own decisions, because you are afraid to take the risks involved in being a manager, they will be demotivated. Employees draw strength from visible evidence of strong supervision and management. Their initiative can be undermined by evidence of weak or vacillating management.

Ibid., p. 131.

Summary

Among classical theories of human behavior in the work environment, McGregor's "Theory X" assumptions emphasize negative aspects of employee behavior; his "Theory Y" suggests that employees do not inherently dislike work and will actually seek responsibility and better performance if encouraged to do so. This latter theory provides a more enlightened, modern approach to motivation.

Similarly, the Supportive Theory of work motivation creates a more effective climate for the development of human potential.

Herzberg's Work Motivation Theory stresses the importance of motivators *in the job itself*, as opposed to such "hygiene" factors as pay, benefits, working conditions, etc.

In the security function, effective job motivators include giving the employee real responsibility; providing opportunity for individual achievement and self-satisfaction; recognition of performance; allowing for and encouraging growth; and creating the opportunity for advancement.

Belittling or "showing up" the employee, public criticism, inattention, favoritism, denial of opportunity, insensitivity, lowering standards of performance, or vacillation in decision-making are all "demotivators" which the effective Security Manager will avoid.

Review Questions

1. Briefly describe the "Theory X" and "Theory Y" assumptions about human behavior.
2. Briefly describe the Autocratic Theory, the Custodial Theory and the Supportive Theory of organizational behavior.
3. According to Herzberg's Work Motivation Theory, what are the motivators that come from the work itself? Why do the job surroundings or "hygiene" factors not lead to work satisfaction?
4. What are some of the ways in which security management can encourage growth among security employees?
5. What are the "Ten Deadly Demotivators"?

Chapter 11

Promotions

Surprisingly, many employees accept mediocrity in management practices as the rule instead of the exception. The selection of new people for the department, the quality of training, departmental disciplinary standards, motivation efforts, the presence or absence of a structured communication capability—all these have a limited impact on the individual security employee, and consequently meet with limited emotional reaction from him.

This is undoubtedly attributable to the fact that people attracted to the security/law enforcement career field tend to be conservative "rugged individualists" with a high degree of self-discipline. They tend to equate what happens around them with what it means to them personally. They take note of the following: an undiscriminating process for the selection of new employees enhances the existing employee's chances for advancement; an ineffective training program provides the really ambitious security employee, who is willing to train himself, an advantage over the less ambitious; strict or erratic disciplinary standards will punish those who are poorly qualified and poorly trained, not those who are well qualified. And so it goes. Rationalization? Probably. The point is that security people are exceptionally tolerant of management practices; they can and do survive the most difficult job conditions with a minimum of complaint—*except* in the area of promotions.

Employees identify very closely with promotions—"There, but for him, go I." It is a truism that the vast majority of employees have an inflated estimation of their ability and worth. Most feel they are under-used, underpaid, and could quite easily do their supervisor's job, and even do it better. Hence, the vertical movement of a peer within the organization is an emotionally charged event, scrutinized with intensity and, unfortunately, with suspicion.

Obviously, then, management's objective in the promotion process is to identify and promote the best qualified candidate, *with resultant general acceptance and approval of the promotion.* That is both a goal and a formidable challenge.

Identifying Promotional Candidates

There is as much excitement among security management personnel in their role in the promotion process as there is among the ranks. The appointment of a new supervisor or promotion of a supervisor to a position of greater responsibility in most cases has a personal effect on existing managers and supervisors. Naturally, they want the best and most effective person moved up.

In identifying candidates, there is a tendency to get mired down in qualities desired or "qualifications" that should not be at issue. Such factors as one's educational achievements, ability to articulate, popularity, the "halo effect" of some recent incident, and length of service should be considered at some later point but should not be the initial qualifying considerations.

There are two basic qualification factors to be considered in selecting candidates for supervisory or managerial responsibility: (1) the employee's track record in job performance, and (2) the anticipated or expected performance in the higher level job. Other factors are peripheral in nature.

That is not to say that college degrees in Security Administration or Criminal Justice, at the Associate, undergraduate or graduate levels, are unimportant. Just the opposite is true. Continued higher education in the field must be indicative of professional growth as well as motivation. But, as important as growth and motivation are, they follow proven performance and anticipated performance as promotional considerations . . . as long as anticipated performance is based on some known qualitative factors.

The Candidate's Track Record

Employees whose service and job performance are rated as "above average" or "above acceptable standards" would constitute the first group of candidates. A rating of "average" performance or "meets acceptable standards" should, as a rule, disqualify the remaining peers. This is because the highly desirable quality of being an "achiever" is reflected in regular performance evaluations of above average. Average workers are doers. Above average workers are achievers; they obviously go above and beyond what is expected. By most employee performance evaluation standards, they "frequently exceed requirements of the job," always through their own initiative. (Incidentally, employees rated as "outstanding" performers are defined as those who "*consistently* exceed job requirements" as opposed to "frequently.") So the achievement-oriented employee meets the first of the two criteria.

Anticipated Performance in the Higher Job

The tentative candidates (the achievers) must now be analyzed, one at a time, as to how they might measure up to or perform against standards of the higher position in question. If the open position requires scheduling employees, for example, those assessing a candidate must look back on his prior performance for evidence of some demonstrated behavior or action indicating that the candidate could indeed schedule subordinates (which would include schedule revisions, emergency scheduling, appropriate degree of flexibility, etc.). It could be that the candidate has actually done some scheduling at the request of his supervisor. In that case, management could properly anticipate or expect good performance in that area if the candidate is promoted.

More often than not, those attempting to qualify candidates for promotion tend to look for disqualifiers in this anticipatory phase. That is to say, they look for duties that, because of past performance, the tentative candidate could be expected *not* to perform competently in the higher assignment.

Disqualifiers may be identified by management or by the candidate himself. For example, otherwise qualified candidates may disqualify themselves because they do not want to travel or move or do not want shift changes. Some may state frankly that they cannot discipline or evaluate others. Self-disqualification makes management's decision easier.

It is important to note that disqualifiers, as such, should be valid for one time only. Any number of things can happen to change circumstances between promotional opportunities. Management should never assume that an earlier disqualifier still stands.

The most difficult task is the projection that the candidate is not able or qualified to perform the higher job. If, for any reason, that disqualification is subject to question or is not the unanimous decision of security management, then the candidate should not be disqualified. He should move on to the actual competition with the other qualified candidates.

What has not been stated, but should be evident, is that the promotional opportunity is common knowledge; it is not a secret. Thus, when the announcement does come, it is no surprise. Most people do not like surprises in the work environment and react negatively to them. In the process of identifying candidates for the promotion, a healthy and open climate about promotions is established.

Selection of the Right Candidate

The best selection process comes in the form of a Promotion Board. The Board, preferably three in number, should always have as members the successful candidate's immediate superior, a person who will be a peer of the successful candidate, and someone from the next higher rank above the successful candidate's superior. If the promotion is for a sergeant's position, for example, the Board would include one sergeant, the lieutenant for whom the new sergeant will work, and the lieutenant's captain. The Board's chairman, as such, would be the lieutenant because the promotion will affect him most directly.

The Board members, except the tie-breaking chairman, should be considered as equals; thus, the captain's will does not prevail because of his rank. The lieutenant, on the other hand, assumes leadership in this case. Leadership includes responsibility because, as stated above, the selectee will be his immediate subordinate, and a very important subordinate at that.

A frequent problem in promotions has been the exclusion of

the promoted person's new superior from the decision. If the lieutenant was not involved in the selection of his new sergeant, he may disagree with (if not resent) the decision, with the result that he may not work at making the new sergeant successful. And it does require effort from the superior to make a newly promoted employee successful. Subordinates, especially those moving for the first time into new and unfamiliar responsibilities, must have direction and leadership from their boss, or they may fail. The truth is, some secret pleasure is taken in such failures. In our example, it is a way for the overlooked lieutenant to strike back at management for not including him in the selection of his own people; it is a way of saying, "I-told-you-so."

On the other hand, if the man who is responsible for the new promotee is involved in and responsible for the selection, he is bound to do all he can to make that selection a successful one.

From the time the appropriate supervisors are asked for possible candidates, through the screening and identifying of those to appear before the Board, to the day of the actual Board interview itself, not more than ten days should pass. Delays and silence about who has been selected to appear and who is to receive the promotion are counterproductive. They breed suspicion.

The Board Interview

Nothing can be done to ease the anxiety of the candidate on the appointed day, and perhaps that small amount of stress is acceptable, as long as it is not purposefully designed into the process. This interview is an important event in an employee's career and life; he will come spruced up as he has rarely appeared before—and more nervous than he has ever been before.

Because some degree of anxiety is normal, the Board should seek to enable the candidate to relax as much as possible after he has been ushered into the room used for this event. Rather than a very formal civil service type arrangement (single chair poised in front of a table behind which sits the "oral" board), the Board members should be arranged in an informal setting of chairs and/or sofa without a table or desk between them and the candidate—similar to a living room or den arrangement.

The same questions, asked the same way, should be put to each of the candidates. General questions should be asked first, such as "Why do you want this promotion?", "Do you feel you are the best candidate for this job and if so, why?", "What special qualities and qualifications do you feel you bring to this job?" Specific job-related questions should follow, along the lines of "What would you do if . . ." and "How would you handle _____ if no one was available to give you direction?"

The answers to the questions will identify the best candidate, and the decision to promote is usually relatively easy.

Following the Selection

When a promotion follows the selection process described, the attitude of those candidates interviewed who did not get the promotion is usually one of full acceptance of the decision and appreciation for the opportunity to compete. The attitude of the organization is one of full acceptance of the promotee because he is the best of the candidates and his selection was in no way political. The attitude of management is one of pleasure and confidence that the best candidate was identified in a totally objective fashion and the newest member of the management team is properly qualified.

Other promotion processes have somewhat less to offer than the board approach.

Promotion from "Within"

Certainly the policy of promoting from within the Security Department should always be followed except when such policy would not serve the best interests of the organization. When would the best interests of the organization not be served under this policy? That situation would arise when an unqualified employee (using the same qualifying criteria outlined earlier in this chapter) is moved up simply to satisfy the "promote-from-within" policy. This type of promotion will destroy, or at least damage, the person promoted; it will automatically impact on departmental performance; and, also of great importance, it will challenge the credibility of the promotional process itself.

If management determines that there is no qualified applicant for a particular post, usually in the higher echelons of the department, then those employees in the rank from which the promotion should come must be advised they fail to qualify for the position and the organization is going to look outside for the necessary talent.

Some employees will initially react against that decision and will ask for specifics as to their deficiencies. They are entitled to such information, which management should know in very specific terms if the candidates were assessed honestly and the disqualifiers were identified. To share the decision openly and to sit down and talk about it with the employees overshadows the initial resentment and paves the way for the future arrival of the selected "outsider."

If the decision is made to go elsewhere for talent and the organization is *not* advised of that decision, the result will be predictable resistance to the new arrival, as well as broad resentment not only for the new man, but for top security management as well.

Vertical Promotion Outside the Department

There are still many in all levels of management who view promotions as organizationally disruptive. They are secure in the status quo. But the "disruptiveness" of vertical movement breathes life, excitement and motivation into the organization. For that reason, promotions—not only upward within the Security Department, but in the company as well—should be explored and encouraged. The advantage, not only of creating promotional opportunities within the organization, but of having good security people in responsible positions throughout the company (such as in personnel administration) should be obvious.

To purposefully hold people back because it serves the immediate purpose of the security organization is morally wrong. And the company as a whole will benefit when there is opportunity for vertical movement both within the security department and outside of it to other areas of the organization.

Advantages of Multiple Layers

Perhaps the best example of the organizational advantages of multiple layers of rank is in the military service. Instead of just three layers of the enlisted ranks—private, corporal and sergeant—military organizations have many layers with, at last count, six graded army sergeant ranks alone. That is true, too, in the officer ranks of the military. The obvious advantage is increased opportunity for vertical movement. The more ranks, the more chances one has to move up. The fewer ranks, the less chance—not only in terms of layers to go up through, but in frequency of openings. Obviously, the greater the number of ranks, the more frequency there is in movement.

The organizational levels in one large retail department store chain are far from typical, but its use of multiple layers may still be of value as an illustration. Starting at the lower levels and proceeding upward are the following ranks: Fitting Room Checker, Fitting Room Inspector, Lead Fitting Room Inspector, Security Agent, Special Agent, Resident Special Agent, Senior Special Agent, Special Agent-in-Charge, (Divisional) Chief Special Agent, Security Manager, and Security Director.

Obviously the small security department cannot have as many levels of rank, but within the limitations of size the opportunity for progressive movement should still exist. Truncated structures limit movement and contribute to stagnation and frustration among the ranks. Rather than few layers with big pay jumps, it is best to have more layers with smaller pay differences and more frequent advances.

Today, in our fast-moving society, people need to feel they too are on the move, and multiple layers help to satisfy that need.

"Temporary" Promotions

An excellent way to measure an employee's potential for higher levels of responsibility is to appoint him temporarily to such posts during natural absences of the regular supervisors or managers— vacations, sickness, leaves of absence, etc. The subordinate's performance while filling in is a measurable indicator of his or her capacity to assume greater responsibility.

Another way to test prospective leaders is to create temporary leadership assignments for special events or projects, appointing one as "team leader" or "project leader" for the duration of the event or project. Not only can they be evaluated, but they also gain the otherwise unavailable experience that comes with increased responsibility. People have been known to excel or to fail under such conditions, and that knowledge can be invaluable to management in future promotion considerations.

"Retreating"

No matter how carefully an organization approaches the promotional process, mistakes in selection are sometimes made. The question is, what do you do when a person is promoted to a position beyond his ability? The "Peter Principle" (rising to one's level of incompetence) is a well-known management problem because of its common occurrence.

There are but three courses of action available to management upon discovery that an employee has been promoted beyond his level of competence:

1. Permit the employee to continue performing in an incompetent fashion.
2. Terminate the incompetent employee.
3. Allow and arrange for a "retreat" back to the former rank.

Option number one, although frequently exercised, is unacceptable to a progressive and enlightened management, for three reasons: it is counterproductive to the organization; it affects the morale of the incompetent's subordinates in a very negative way; and, finally (a point often overlooked), the person who has achieved the level beyond his ability knows it as well as anyone else, and that knowledge places him in a dilemma. He knows he is in trouble but does not want to admit it. He struggles. Unhappiness sets in. Some actually become physically ill because of the dilemma. It is a rare individual who can come forward and admit he is in trouble. The tragedy is that most will not admit it.

Option number two is another popular solution. The tragedy here is that management is culpable, too. The man promoted beyond his ability was certainly competent at the lower rank. In fact, his skill level and performance evaluation were above average. He was a good worker. He is not inherently an ineffective employee—he is simply ineffective in his new responsibility. To terminate this employee is in part to hide management's mistake in promoting the wrong person.

Option number three, if it can be accomplished, serves the best interests of all concerned. Policies prohibiting voluntary demotions are unrealistic and inhumane. Certainly the man who is willing to retreat to his former rank should be given that opportunity, even though his ego is going to be bruised for a time. The total honesty in "retreat" situations is apparent to all observers. Not only is the salvaging and protection of a man's tenure with the firm important to the misplaced individual, but others in the organization also respond to this humane policy. The first two options are known to the organization; employees have a collective wisdom of some dimension.

"Retreat" should never be a structured or formalized consideration in the promotion process because, by all rights, it should be the tragic exception in organizational life. To say to each candidate, "Well, if you prove incompetent, Harry, you can always go back to the old job," would discredit the selection ability of the Board (or management), would demean the importance of striving for competence by overcoming obstacles, would turn promotions into a gamble instead of a challenge, and would be an insult to the real achiever's confidence and intelligence.

"Retreating" is the emergency valve that should be available for those rare instances when the promotional process fails to hit the mark.

Summary

Because vertical movement in the organization is an emotionally charged event, management's objective in promotion is always to identify and select the best qualified candidate.

Primary qualification factors in promotions are the employee's record of performance in his present job and his projected performance in the advanced position. Current job performance should be above average to be considered for promotion.

A Promotion Board provides the selection process most likely to be fair and objective. In all cases, the supervisor for whom the candidate will be working should be a member of the Promotion Board.

The opportunity for vertical movement elsewhere within the company should not be opposed by security management. Such movement serves both the company and the security organization.

The most effective promotion process is a flexible one. It will not make promotion automatic, but will promote only the qualified candidate; it will accept the necessity to go outside the department in some circumstances to find a qualified employee; and it will also allow for the possibility of "retreat" where an employee proves out of his depth in a higher level position.

Review Questions

1. What is management's objective in the promotion process?
2. What are the two basic qualification factors to be considered in selecting candidates for supervisory responsibility?
3. What problems can occur if the promoted person's new superior is not involved in the selection decision?
4. Discuss the advantages of having multiple ranks of employees.
5. Discuss the relative merits of the three possible courses of action when an employee has been promoted to a position beyond his ability.

Chapter 12

Communications

Of all the qualities, talents and skills required of a manager, the art of effective communication must rank first. In fact, all other managerial factors are crystallized by the communication process to form the "whole" administrator and leader. Stated another way, the manager who is a strong communicator is a strong manager, and the manager who is a weak communicator is a weak manager.

Consider this: the manager's task is to get others to do the job *when* he wants it done, *how* he wants it done, *where* he wants it done, and (if he is a good manager) to make them understand *why* he wants it done. Obviously, to translate his desire into the completed task, the manager must communicate the desire, and the employees must understand. If a job is done poorly, not done at all, or done incorrectly, it is usually because the employees *did not understand*. The manager failed, somehow, to communicate clearly.

The example above reflects only one type of communication—verbal—down. Other types of communication to be considered include: verbal—up; written—down; written—horizontal; written—up; verbal—horizontal; and action. These types of communication may take place in an "Open Climate of Communication" or a "Closed Climate of Communication."

Verbal—Down

Traditionally, security organizations generally operate under the Closed Climate of Communication. This is probably due to the quasi-military nature of most security forces. It is easy to visualize the Captain on the bridge of a Navy vessel shouting directions into a tube that terminates deep down in the engine room, and everyone complying with the orders. This type of communication might be called "stovepipe" communication (verbal—down only). "Stovepipe" communication may also occur when the Captain of the Guards passes word down to the officer to lock Gate 36 at 2:00 a.m. As a rule the gate will in fact be locked as required. However effective this method of communication seems, there is an obvious flaw which haunts managers (or Captains): sometimes the gate will be found unlocked. The officer at the bottom end of the "stovepipe" hears the orders, but there he stands silently without the right key!

Ideally, then, management should strive for the development of an Open Climate of Communication.

Verbal—Up

Some readers may feel that the above example, where the officer at the bottom end of the "stovepipe" hears the order, knows he does not have the key, but remains silent, is unreal. All the officer has to do is tell the superior closest to him on the "stovepipe," probably the sergeant, that he does not have the key. This is not likely to happen, however, because the officer knows that if he tells the sergeant he does not have the key, he will be reprimanded for forgetting it or be embarrassed or abused in some other manner. To protect himself he remains silent, knowing the job will not be done. Rather than feeling free and undefensive and openly communicating to his supervisor (verbal—up) so as to get the job done, he voluntarily chooses to undercut the effectiveness of the organization—all because of the Closed Climate of Communication.

In most cases the organization does not set out to purposefully design a closed climate of communication. The climate is created at the top (the Security Director or Security Manager level) by insensitivity to the importance of communication, including verbal—up.

Written—Down

There is a classic exercise or game which dramatically points out the unreliability of spoken communications. A group of people are seated around a room. Instructions, or a short story, are whispered to the first person. He, in turn, repeats in a whisper what he heard to the next person, and so on. It is hilarious to hear the last person in that room speak out loud what he or she was told—but the reality which this game illustrates can be tragic as well, because, in any organization, many failures are attributable to the misunderstanding of directions.

There are many factors in the breakdown of verbal communications:

- People tend to hear what they want to hear.
- Generalizations are narrowed to specifics.
- Different words mean different things to different people.
- The spoken emphasis on a word or phrase can be interpreted differently by the listener, so that he assumes a different meaning than the speaker intended.

It stands to reason that there is less chance for error in understanding the written word. Not only is there less chance for error, there is more acceptance and credibility in the written word, especially in organizational life. How often does one hear, "I want to see it in writing"?

Logically, then, everything that can be reduced to writing should be. General orders, post orders, patrol orders, inspection instructions, rules and regulations which are to be enforced, rules and regulations governing the Security Department, investigative procedures, emergency procedures, alarm procedures and other appropriate material and data can and should be in writing and available to members of the department.

Should all such written material be distributed to all security personnel? Probably not. All written material will fall into three categories: "Nice to Know," "Should Know," and "Need to Know." Certainly every member of the department should be provided with the "Should Know" and "Need to Know" material.

Written material should always be put into a standardized format. For example, Orders of the Day should always be on the same size and color of paper, in the same print, with the same distinctive appearance. In this way it is recognizable at a distance as an Order of the Day. The same should be true of all other written material. Standardized formats reduce confusion, make for easier organization, and give a professional touch to the department.

Department Handbook. The need for a departmental "How-to" handbook should be apparent to any Security Manager or Director. Even a small security organization typically faces the problem of relatively high turnover of personnel as well as the need for changing or rotating assignments. And as the department grows and specific roles become more specialized, the need for standardized procedures becomes even more essential. One store chain's handbook, or manual, contains an assortment of detailed instructional materials, such as a standardized confession format, investigative steps to follow in tracking a "shopping violation," and the proper procedure to follow in processing restitution from an employee's terminal wages.

The manual need not be fancy or pretentious. Many organizations use a loose-leaf format, making it easy to add or change pages as needs change.

What is important is to develop a departmental handbook that will be a truly functional tool—one that will be *used,* providing quick instruction for the new employee or the guard assigned to a new post, spelling out procedures to be followed in any given situation, serving as a source of authority to settle disagreements or resolve confusion—especially valuable where there is wide geographic dispersal of personnel, such as in a large manufacturing plant or a chain of stores.

Ideally, the security handbook or manual should also be part of the company's general managers' manual. In this way not only will security people be provided with written material about their organization and job, but so will management. Everyone will be speaking the same language.

Departmental Newsletter. In an organization with a closed climate of communication, there is a reluctance to reduce policies and procedures to writing, or, if they are put down on paper, a reluctance to make such material available to the employees.

The "written—down" type of communication goes beyond policies, procedures and how-to's. A departmental newspaper or newsletter is an excellent communication medium. In one security organization, *The Rap Sheet* is an eight- to twelve-page monthly written for the general interest of Security Department employees. A typical issue will carry a general motivational or good management techniques statement from the Director; a "Security Officer of the Month" article, with picture; a listing of promotions and transfers; an "In Response to You" column (sort of a "Dear Abby" column answered by the Security Director); pending or new laws that will have an impact on the organization; interesting arrests and/or investigations; interesting security statistics such as arrests for the year to date against last year's figures; a security question of the month, with the answer to the previous month's question; a security-oriented puzzle, and perhaps a security cartoon.

The benefits of this type of communication are innumerable. Employees more closely identify with the organization and feel a part of the department. They appreciate being kept abreast of the latest happenings; they are, in fact, well-informed. They love the recognition afforded them in print. The Director has a vehicle to make known his standards and goals. In the eyes of the company and its non-security employees, the polished security publication is another indicator of the professionalism growing in the industry today.

Written—Horizontal

An example of written—horizontal communications is a company-wide security newsletter, providing the security administrator with the magnificent opportunity to communicate what is happening in the security world to the company as a whole. Too often, security departments are considered mysterious and organizationally non-contributory functions. The good security administrator can bring recognition to his department by opening up communications and sharing information with management of the company whenever possible.

As an example, one retail security organization publishes a monthly *Security Newsletter for Management*. The objective of this publication is to make management personnel aware of the security risks and security achievements in the industry. This simple, clean-

looking four-page newsletter opens with a series of condensed incidents of recent occurrence in the retail community at large. One
typical incident:

> In a competitor's store in the downtown area, four adults
> of Latin descent, two male and two female, entered the women's
> sportswear area. While one of the females engaged the sales-
> woman in a complicated transaction, her companions removed
> a total of ten fur-trimmed suede coats from the racks. Tech-
> niques used: the woman rolled four coats and crotched them
> (placed them under her dress and held them between her
> thighs), and each of the men wrapped three coats around his
> mid-section under a raincoat. All four escaped in a white Falcon
> station wagon bearing out-of-state license plates. Total loss:
> $1,700.

That story is followed by others, still concentrating on attacks
against competitor stores. These might include incidents of malicious
mischief (juveniles setting off sprinkler heads, with resultant water
damage), use of stolen credit cards, etc.

Following these are incidents within the company, chosen for
dramatic impact. Again, the objective is to make management aware
of security problems and the Security Department's efforts and successes in these problem areas of the business.

Among the incidents described are stories of recent employee
dishonesty cases (names of employees are omitted). Each of these
vignettes concludes with information on the disposition of the incident, such as termination of an employee or police department
booking. Figure 12–1 illustrates a typical page from the newsletter.

Unit managers not only read each and every word of each publication, but they circulate the newsletter to their staff members and
then read it aloud at departmental manager meetings. Discussions
follow. The monthly impact remains constant: renewed amazement
at the scope of the security problems; amazement over cleverness
in tactics; shock at the constancy of internal theft problems.

The end result is greater security awareness within the organization.

Written—Up

In recognition of the need for employees to communicate upwards (and the need is as much for the Director to know what is

DISHONEST EMPLOYEE	Region I
	A member of the Housekeeping staff was observed leaving store with a bucket and mop on his way to clean the Tire Center. His bucket was checked and under the mop was a Craig Stereo cassette tape deck and a Sony AM/FM radio. Employee was escorted back to store where he subsequently admitted theft and was terminated.
DISHONEST EMPLOYEE HANDOUT	Region I
	An employee in Jr. World had a friend come into department. The friend selected a top and a pair of pants and went into the fitting room. Once her friend had entered the bank of fitting rooms, the employee stood at the entrance and acted as a lookout. When the accomplice emerged from the fitting room she had only the pair of pants. After a brief conversation between the two, the accomplice left the store. The accomplice was later arrested in one of the Mall stores where she worked. The employee was terminated and both were booked by the Culver City Police Department.
ATTEMPTED GRAB & RUN	West Covina
	Night Service Manager, Mr. Smith, prevented a "grab and run." Smith observed a car parked outside the store's door in an unauthorized parking space and took down the license plate number. He then waited around to see what the two males in the car were waiting for. He then observed a male grab an armful of Dept. 50 merchandise. Smith then took after the suspect, whereupon the suspect turned around and threw the merchandise at Smith.
	License Plate Number:
	Total Recovery: $473.00
SHOPLIFTER	Fashion Valley
	A Roving Security Officer observed a F/C, 5'6", 115 lbs., 34 yrs. old, blonde curly hair and glasses, carrying a stuffed handbag clutched tightly to her side. The "customer" exited the store and came back shortly with an empty handbag. The R.S.O. followed the "customer" to the Children's Department where she rolled a child's robe and gown and placed them in her handbag. On her way out, she stopped in Cosmetics and picked up some lotion which she also put in her purse. The "customer" was apprehended as she exited the store.
	During interrogation she confessed to having made several trips into the as well as other mall stores. Over $600.00 worth of merchandise was recovered...$500.00 from and the remainder from and . She was arrested and booked by the San Diego Police Department for Grand Theft.

Figure 12-1. Typical page of a Security Newsletter for Management.

on the minds of his subordinates as it is for every employee to have the opportunity to express himself), *The Rap Sheet* already mentioned was originally designed as a two-way communication tool: written—down and written—up.

How was this accomplished? The first, opening edition of *The Rap Sheet* stated that, in an effort to further open communications, questions, suggestions and complaints were solicited from all security employees, with the promise that *every* such question, suggestion or complaint would be answered. The response was gratifying.

Not all suggestions are adopted. Not all problems are solved. But all receive response. All employees have a way, without putting themselves in jeopardy, to have their say, to be heard by the Director and to hear what he has to say in response.

If an Open Climate of Communication is to be established, some such two-way avenue of expression is essential.

Verbal—Horizontal

There are two types of verbal—horizontal communications within the context of our definition of an Open Climate of Communication. The first is intradepartmental (i.e., security personnel only) and the second involves communications with other departments in the company.

"Rap sessions" constitute the interdepartmental type. One such session might have given sub-units of the Security Department, such as the Fraud Unit, sit down together without regard to rank and talk about their work—the practices, techniques, problems, failures and successes—with no specific objective in mind except to communicate. As a rule, something of value will surface in these sessions. This could be anything from a clarification of a misunderstanding between two peers to a more realistic deadline on certain types of cases. The important result is that all participants leave the session with a good feeling about themselves, their unit, and the Security Department as a whole. As individuals, they had a chance to be heard, a chance to think out loud, a chance to be themselves.

The second type of horizontal verbal communication occurs when the Security Director and any number of his staff people go out into the company and meet with various units. These meetings are also "sit-down-around-the-table-and-talk" sessions. In these settings questions are encouraged and Security speaks openly and honestly to the questions.

The benefits of broader communications within the company are invaluable. Questions such as, "What right does Security have in searching our briefcases and parcels when we leave the building?" give Security the opportunity to cite the authority and talk about the *why's*. Perhaps not every person will be satisfied with the reasons, but they will leave such meetings with more understanding and appreciation for security. This assignment calls for a security representative who is comfortable and at ease in this type of challenging environment. How he handles questions and complaints can leave a very favorable impression on participants in attendance— or a very negative impression.

Communication, then, is the very lubricant that makes the managerial machinery run smoothly and efficiently.

Action

It is true that action speaks louder than words. The administrator who wants to establish a climate of open communications had better be prepared both to listen and to respond.

If management is sincere in wanting suggestions from employees about the running of the Security Department, it stands to reason that some of their ideas will have merit. In fact, the Security Manager will find it a constant source of amazement to see how smart and creative so many people are. Some managers feel that "I am the boss, and I should have all the workable ideas!" This attitude is shortsighted, but unfortunately a not too uncommon managerial disease.

So, if you ask for ideas, you must adopt some—the ones that are meaningful and contribute to the success of the organization. If you ask for complaints, then you must be prepared to take appropriate corrective action to cure those complaints.

If you reject sound ideas, you will discover over a period of time that the source of ideas within the organization has dried up; there will be no more upward communication. If you are critical of questions or are unresponsive or evasive, the questioning will taper away to nothing; no more upward communication. And if you disregard the message contained in complaints and fail to react in a positive and corrective way, you will lose the benefit of hearing what is troubling personnel. An example in this connection would be the case where the Director hears from a number of employees that a certain supervisor engages in heavy-handed supervisory techniques and intimidates his subordinates. If the Director disregards such information and takes no action, or even promotes the supervisor in question, the credibility of his Open Climate of Communication becomes a joke. If, on the other hand, the Director causes the supervisor to be exposed to a leadership training program, his credibility and his communication "program" are maintained.

Summary

Communication ranks at the top of the effective manager's skills. Both the organization and its employees will be better served by an Open Climate of Communication, both upward and downward, whether verbal or written.

Better understanding is assured when communications are in writing wherever possible. Written-down communications might include the essential Department Handbook as well as departmental newsletters. Avenues should also be provided for *written-up* communications, in which the employees have the opportunity to express their ideas and feelings to management.

Horizontal communications, both within the department and between Security and other employees and units of the company, are mutually beneficial. And an Open Climate of Communication will remain viable and credible only when management listens—and responds.

Review Questions

1. What is meant by "stovepipe" communication? Give an example of the shortcomings of this type of communication.
2. Discuss several factors in the breakdown of verbal communications.

3. Briefly describe two types of newsletters that might be effective communications tools for the Security Department.
4. Discuss the ways in which a security manager can establish an Open Climate of Communications in his department.
5. Describe two types of "horizontal" communications and their potential benefits.

Chapter 13

Career vs. Non-Career Personnel

The question of utilizing career or non-career personnel to discharge the security function within a given organization, with the respective advantages and disadvantages of each, is worthy of examination.

"Career" personnel may be defined as full-time (usually on a forty-hour work week) in-house employees on a "career path," with apparent intentions and aspirations to grow in the organization, in the security career field or in another career field within the company (such as personnel services). "Career path" means, in this context, a track leading to continual vertical movement within the greater organizational pyramid.

There are two major categories of "non-career" personnel: (1) in-house part-time employees with a short work week such as twenty hours; (2) employees of another company who perform duties for the organization on a contractual or service fee basis.

At the outset it must be understood that many factors have an impact on the type of personnel to be used in the security function. The most important factor is the character and nature of the function itself. What could very well be an advantage in one security operation might be a distinct disadvantage in another. Yet, the following somewhat generalized categorization should provide some insight into the differences between career and non-career personnel.

147

The title of this chapter may suggest that an axiomatic choice of one approach over the other will be the ultimate result of the comparison, but that is not the case. Rather, an objective consideration will reveal good points and bad points, advantages and disadvantages, to both career and non-career personnel; and in view of that, the decision to have a *blend* of both might well be the best solution to the security manpower needs of a particular organization.

CAREER PERSONNEL

Advantages of Career Personnel

1. Companies with in-house security programs tend to attract people seeking career positions and career opportunities. Competition for such openings allows selection of the most qualified individuals. Attractions of career jobs include salary, pension or retirement plans, profit sharing programs, the entire employee "benefit package" (which usually includes medical, hospitalization, dental and life insurance, vacation, and sick leave), and employee privileges unique to the company (such as merchandise discounts in retailing and free or near-free travel for those in the transportation industry).

2. Career personnel develop a loyalty to the department as well as to the company. They identify with the organization and see its welfare as synonymous with their own.

3. Career personnel tend to have greater knowledge of the company, its "ins and outs," and with such insight they function more efficiently and smoothly.

4. Career personnel establish an *esprit de corps* or comradeship, and the resultant pride reflects in their performance.

5. Career personnel tend to be more ambitious and motivated to work due to apparent opportunities for advancement.

6. There is more stability among career personnel in terms of turnover in an organization, primarily because of seniority and vested interest.

7. There is more communication between the security function and the rest of the company when career people are in place, usually because of mutual company identity and common company interests.

8. Career personnel tend to be better trained because training costs are "hidden"; that is, the cost of training is part of salary

expense and is not identifiable as an extra expenditure. In the case of contractual service, the time which the contract personnel must spend in on-the-job training with the client company is clearly unproductive.

9. There is a higher degree of technical proficiency among career people because the company is willing to invest the necessary time and money to train them in anticipation of pay-back through long-term service.

Example of the Advantage of Career Personnel: Telephone companies are an excellent example of the many security organizations that utilize career personnel almost exclusively. Telephone company security agents identify closely with their firm, have in-depth knowledge of the telephone communication business, and take pride in their department as well as their employer. They rarely change companies. They know people throughout the company, because in most cases they came from non-security ranks. They are well trained, and, finally, have a high degree of technical proficiency.

Disadvantages of Career Personnel

1. In terms of costs, career personnel are substantially more expensive than non-career people. For example, there is an additional cost of about thirty percent above and beyond the payroll costs that supports the employee benefit package. Other costs, either capital or sundry, including everything from equipment and office facilities to uniforms (to name but a few), are unavoidable in a career organization. Appreciable savings in this area can usually be realized in contractual agreements because these costs have already been incurred by the contractor.

2. Career personnel constitute a fixed, limited cadre or pool of manpower resources. Special events, special problems or emergencies could well sap the organization and have an adverse impact on the daily security requirements.

3. There is a certain amount of inflexibility in the deployment of career people in terms of location and time scheduling, more frequently than not due to company policy. Personnel policy might require three, five, or seven days' notice of a shift change, which obviously limits security management's flexibility in its attempt to provide protection. Career employees often enjoy "portal to portal"

pay and travel time allowance for reporting to a location other than their regular place of work, whereas contract services might have personnel already in place at the distant location, with no loss of time or additional expense.

4. Because of the employee-employer relationship between the career employee and his company, certain disciplinary restrictions can be departmentally counterproductive. The inherent obligation of management to its employees, the source of which is traceable to governmental, administrative and judicial rulings, affords every employee job security—to the point that even those employees who, for one reason or another, perform at a marginal if not substandard level, must be retained for lengthy periods of time prior to their discharge. There is no such employee-employer relationship with non-career people in place.

5. Career manpower resources have limited parameters and ceilings of talent, and departmental capabilities are restricted as a result.

6. There is the ever-present problem of those career people who "top-out" at one level or position, and upon being told or otherwise realizing they will not progress any further, become disenchanted or resentful. Such employees frequently will not leave, and even though there is an attitude problem, management cannot terminate them because their job performance meets standards, however minimally.

Example of the Disadvantage of Career Personnel: Personnel policies of many organizations require a series of job performance "cautions" and warnings, alerting the employee that he is performing below standards and giving him the opportunity to improve. For example, one company requires three such warnings, spaced at least thirty days apart, *before* the employee can be given notice. Thus the department is obliged to endure approximately 120 days or one-third of a year of substandard work.

Additionally, an individual so terminated may file a legal action against the company on the grounds that the company's entire action against him was not based on work performance but on one form or another of prejudice (race, color, creed, sex or age). The company must then assume a legal defensive posture (at no small expense) and subsequently may be obliged to reinstate the former employee *with full back wages* for all time elapsed between termination and the final determination of the issue—which could constitute a full year's wages all at once.

PART-TIME NON-CAREER PERSONNEL

Advantages of Part-Time Non-Career Personnel

1. Part-time in-house employees are less costly than career personnel because they are not entitled to the full employee benefit package.

2. Part-time employees, by virtue of the agreement made at the time of hire regarding their schedule, allow for security coverage at difficult, unusual or odd hours that otherwise could require overtime or premium pay to regulars. This allows for broad and flexible coverage.

3. Deployment of part-time people permits security management the unique opportunity to analyze their performance over an extended period of time. If a part-timer proves productive he can move into an unfilled career position if he desires that opportunity.

4. The use of part-time security employees allows management to tap particularly high-caliber people for security service. Local colleges and universities are an excellent recruiting ground for parttimers. Intelligent and capable college students can make a substantial contribution to the security function in a wide variety of capacities. Part-time schedules usually fit into their school schedules as well as their financial needs, and this employment proves to be a bargain both ways.

Also, more and more women are returning to the labor market, many of whom have raised their children and wish to be productive members of the community but do not want or cannot handle career or otherwise full-time jobs. These people bring maturity and common sense to part-time positions.

Finally, a number of people who have retired early still need to be productive, and they, too, can be a definite asset to the Security Department.

5. Off-duty police officers can be a good source of part-time security employees because of their availability, interest in some phases of security work (those assignments which most closely parallel their own profession), and their comprehension of the protective

function in general. Off-duty police officers possess trained talent and steadfast loyalty to the ideals of law and order which underpin security's existence. They have been fully screened for integrity and honesty.

Example of the Advantage of Part-Time Non-Career Personnel: Providing security coverage for an amusement park that is open twelve hours a day would pose a real problem to the security administrator responsible for such coverage—if, that is, all security personnel were career people. The schedule would call for X number of people to open the park and X number to close the park the same day. Thus there would be an unproductive (and wasteful) expenditure of overlapping hours and payroll dollars.

On the other hand, if part-time security personnel were deployed along with full-time regulars, full coverage could be effected with economy. For example: ten officers are required to open and ten to close. The park is open from 12 noon to 12 midnight. With full-time employees only, the first ten officers go on duty at noon and go off duty at 8:00 p.m. The second shift must go on duty at 4:00 p.m. and work to midnight to receive their full eight-hour shift. Between 4:00 and 8:00 p.m. there are 80 manhours being expended when only 40 are required. One full-time unit is wasted each day.

With effective scheduling of part-time personnel, the coverage could be as follows: five regulars and five part-timers (working a four-hour schedule) start at 12 noon. At 4:00 p.m. the part-timers go off duty and are replaced with five regulars who will work until midnight. At 8:00 p.m. the original five regulars go off duty, replaced with five part-timers who work until midnight. The schedule of coverage is economical and efficient.

Without question, utilization of part-time personnel helps alleviate scheduling problems.

Disadvantages of Part-Time Non-Career Personnel

1. Part-time employees have a decidedly limited commitment to the job and to the organization. They do not feel the same degree of responsibility as does a career employee. The limited feeling of responsibility results in more part-time employees failing to report, calling in sick, or offering other excuses for not appearing. Therefore, part-timers are less reliable.

2. The primary interest or attention of part-time personnel is somewhere other than the job—school, family or another job. Consequently, keen interest and attention are usually lacking. They do not identify with the company, and as a result their conduct on the job is affected adversely.

3. The relationship between the company and the part-timer is essentially mercenary in nature: immediate remuneration for services rendered. That means the primary work motivator of the part-timer is money, not achievement, challenge, growth or responsibility.

4. There is a limited number of people looking for part-time work. By far most people in the job market want full-time employment.

5. By virtue of their own emergency status in the public sector, off-duty police officers who work part-time in security cannot be counted on by the company when a major disturbance or calamity occurs in the area—those very times when they are needed the most.

6. Another disadvantage of the off-duty police officer as a part-time security employee is his tendency to take more chances, because of his experience and peace officer status, than would a civilian, particularly in making arrests.

Example of the Disadvantage of Part-Time Non-Career Personnel: A department store plans coverage of a given store from opening to closing, using a part-time employee in the plan for evening protection. Because of absence of a real commitment, the part-timer fails to show for any number of reasons, leaving the store short of security help. The question of dependability is the biggest disadvantage in the use of this type of employee.

CONTRACTUAL NON-CAREER PERSONNEL

Advantages of Contractual Non-Career Personnel

1. There is a considerable cost savings, in terms of the expense of the employee benefit package and other career employee privileges, when contractual personnel from outside the organization are engaged.

2. There is complete freedom to terminate the services of an individual serving the company on a contractual basis. Such termination can be immediate and without cause. That means that if his

appearance, grooming, attitude, age, demeanor or performance is for any reason below the standards set by the company, the person can be removed from the job and returned to *his* employer, without repercussions.

3. There is good flexibility in manpower resources in a contractual arrangement. The security force can be increased to meet unexpected demands overnight if need be. This can be achieved by the primary contractor sending more personnel or calling on another contractor for short-term assistance.

4. Use of contracted services reduces miscellaneous non-security expenses such as recruiting and advertising costs, personnel interviewing and administration costs, timekeeping and payroll administration costs.

5. There is freedom to terminate services of a contractual firm if that firm's services fall below required standards. Contracts, whether written or agreed upon verbally, usually allow for a thirty-day termination clause, and poor performance justifies execution of this clause.

6. Flexibility of coverage and service in a geographically dispersed operation is a decided advantage of contractual help.

7. The short-term and/or infrequent needs for personnel with unique or highly specialized skills and technical know-how, such as a polygraph operator, can effectively be met on a contractual basis.

Example of the Advantage of Contractual Non-Career Personnel: Consider a case where the Security Department causes the placement of an undercover agent into a warehouse for the purpose of gathering information on possible internal theft. The undercover agent's *primary* employer is a contract service firm. He receives a salary from them as well as a regular paycheck, like every other warehouse employee, from the company that owns the warehouse. For a period of time some useful intelligence is obtained, but after a while the undercover agent becomes personally involved with other warehouse employees and his reports become valueless. Even though he wishes to remain employed in the warehouse, his services can be terminated forthwith without violating his rights to job security, because his real employer is the firm which sent him to the warehouse and is still paying his undercover salary (although it may be less than the warehouse salary).

If, on the other hand, the Security Department had hired an

applicant directly into the warehouse to serve as an undercover agent, that person would be entitled to some job protection and could not be summarily removed from the job. The use of contractual services has some very definite advantages.

Disadvantages of Contractual Non-Career Personnel

1. Except for those few firms that pay excellent salaries, most contractual firms attract personnel with relatively poor qualifications. In order for the contract firm to be competitive and make a fair profit, the individual must settle for a lesser wage than would be paid for a comparable job in a non-contractual firm.

2. There is more turnover in non-career personnel because they may find what they believe is a career or a better job elsewhere, or because their talent can best be used by the contract firm to the firm's advantage. That means, of course, that the highest rate of turnover occurs among the most talented people sent to service the client.

3. There tends to be an absence of pride among contractual service people—pride in themselves, in many cases, as well as pride in their organization—and the absence of pride reflects in performance.

4. There tends to be a resentment on the part of many contract people over the fact they work for a company (client) but are not entitled to the benefits the regular employees receive. As a result, some contract employees seek to be hired by the client company as regular employees. Most client-contractor agreements now include provisions prohibiting such job changing or requiring the payment of a fee by the client if he wants that employee. (Note: These agreements only validate the point that contract people frequently would prefer to identify with the other employees and the client's firm.)

5. Ambition and motivation are questionable in many cases because opportunities for advancement seem limited, or the contract firms have failed to lay out meaningful and comprehensible career paths and make their people aware of such opportunities.

Example of the Disadvantage of Contractual Non-Career Personnel: A major shopping center contracted for guard services, awarding the contract to a reputable firm with excellent leadership at the top. The securing of this particular contract was an important addition to the list of clients being served, and good people were assigned to this most visible job. As time passed, the contract firm grew and turnover increased. The guards on the site became careless of appearance, inattentive, unreliable and eventually became a source of embarrassment to the shopping center. Such a condition could not have developed had the service been proprietary, with career people in place. (On the other hand, some shopping centers, to name but one of hundreds of types of clients, have had excellent results with contract services.)

Combining Career and Non-Career Personnel

Clearly, then, there are many factors to be considered in weighing the pros and cons of career and non-career personnel. In addition to those factors discussed above, there are others not explored, but also most important. Those are reflected best in simple good management practices and supervisory skills. Despite their disadvantages, career personnel can excel, depending upon good management. Non-career personnel, contractual or part-time, despite the disadvantages enumerated above, can excel if given proper supervision and good management.

Early in this chapter reference was made to the possible utilization of both career and non-career people—a blend of both. That is precisely the practice in many organizations. In one retail chain the core of the Security Department is comprised of well-trained professional career personnel. Complementing them is a large cadre of part-time security employees, including a large number of college students, many of whom are studying in the criminal justice curriculum. At the time of this writing, supplemental contractual services include some uniformed guards, undercover agents, temporary "agent" assignments, and integrity shoppers. To provide adequate protection with career people exclusively would be an impossibility. To protect the company with non-career personnel exclusively would likewise be an impossibility.

An appropriate balance or blend of both is recommended. There is a need for both and room for both in the security industry.

Summary

There are advantages and disadvantages to employing full-time career security personnel or non-career employees drawn from part-time workers or contract security services.

Stability, loyalty, improved "local knowledge," superior pride and motivation, and the opportunity for better communications and training are advantages of career personnel. On the other side of the coin are increased costs, limited numbers and the attendant inflexibility of deployment, and potential problems of discipline and limited skill levels.

Part-timers are less costly; provide desirable flexibility in assignment; and allow management to draw from high-caliber sources such as students and women available only on a part-time basis. However, part-time personnel tend to lack the career employee's commitment to job and company.

Contract services offer the benefits of cost savings, freedom to terminate services at any time, great flexibility both in manpower resources and in coverage of widely dispersed operations, and specialized skills. Commonly cited disadvantages include low-paid personnel, subject to high turnover, with a lower level of pride and motivation.

No one type of employee is right for all situations; in fact, many companies can best be served by an appropriate blend of full- and part-time, in-house and contract security personnel.

Review Questions

1. What is the definition of "career personnel"?
2. What are the two categories of non-career personnel?
3. List six of the advantages of using career personnel in the Security Department. Contrast these with six disadvantages.
4. Give an example of how utilization of part-time personnel can help solve scheduling problems.
5. Describe the advantages, in terms of flexibility, of using contractual personnel.
6. Discuss the statement, "An appropriate balance or blend of career and non-career personnel is recommended."

Part III
OPERATIONAL MANAGEMENT

Chapter 14

Planning and Budgeting

The budgeting process might best be approached in terms of
   WHAT is a budget?
   WHY do we have a budget?
   WHEN is a budget prepared?
   WHO participates in the budgeting process?
   HOW is the budget prepared?
Such a pragmatic approach overrides the broader spectrum of
concepts, philosophies and strategies, such as "zero-based budget-
ing," about which entire texts have been written. The emphasis
here is on fundamentals—an understanding of the basics provides a
groundwork for sophistication and growth.

The established security manager, moreover, is already involved
in an ongoing budget program in his organization. Even the new
manager, whether promoted from within or brought in from outside the company, will inherit budgeting responsibilities in an existing framework. A pragmatic approach to the budgeting process,
therefore, is most useful.

WHAT IS A BUDGET?

The management process is the coordination and integration of all resources to accomplish organizational objectives. According to this definition, management is viewed in terms of the functions a manager performs—i.e., planning, decision-making, organizing, directing and controlling. Each of these functions has an impact, to one degree or another, on the budgeting process. Controlling, for example, is that process aimed at insuring, through overt, timely action, that *events conform to plans.* Plans must be based on good judgment and good decision-making estimates about the future. *The budget is that plan stated in financial terms.* Planning and budgeting go hand-in-hand. You cannot have a budget without a plan; and every plan, if it is viable and is to be executed, must have a budget. A budget, therefore, is:

- a plan stated in financial terms.
- a realistic estimate of the resources required to implement a plan.
- an allocation of resources to achieve planned objectives.
- an instrument which records work programs in terms of appropriations needed to place them into effect.
- a management tool intended to insure that work programs are carried out as planned.

Obviously, the definition of the budget must include plans (or programs, which in and of themselves are plans).

The elements of a budget can be illustrated in a practical security situation. The Security Department decides to provide a rape prevention program for the female employees of the company. The objective of the program is obviously aimed at educating employees on ways to protect themselves against the possibility of rape, ways to increase their safety during movement to and from home, and so on. With the objective established, next comes the planning of how to achieve this educational goal. The plans could include the rental or purchase of a commercially prepared film on rape prevention, preparation of posters announcing the program, scheduling of a security officer's time to conduct the program, rental of a projector to show the film, retaining an outside speaker who is considered an expert on the subject, and distribution of an anti-rape booklet to the participants following the program.

Once the plan spells out what must happen to achieve the stated objective, it must then be costed-out, or restated in dollars and cents.

1. Film purchase $285.00
2. Posters
        artwork................................. 77.00
        printing 34.50
3. Rental of 16mm projector..................... 20.00
4. Guest speaker fee........................... 200.00
5. 300 booklets @ .28 .... 84.00

Rape Prevention Budget $700.50

To repeat, then, a budget is a plan stated in financial terms; it is a realistic estimate of the resources required to implement a plan; it is an allocation of resources to achieve planned objectives; it is a way in which we record programs in terms of the dollars needed to place such programs into effect; and it is a tool intended to insure that the program comes off as planned.

WHY DO WE HAVE A BUDGET?

The budget breathes life into a plan and gives the plan direction. It requires the manager to *manage* the plan in three dimensions:

1. The operation or project must unfold *as planned* if the budget is followed exactly. (If we planned on a film for the Rape Prevention Program, the film will in fact be used if the allocated dollars are spent to purchase the film. Without the budget as a guide, something could easily be substituted for the film, and thus the plan would not be followed.)
2. The operation or project will take place *when planned* because budgets are time-phased; i.e., plans must be executed in keeping with the budgeted availability of funds. In other words, if the salary budget for a six-month period amounts to $600,000, that money is not available in one lump sum at the beginning of the six-month period, but rather is rationed out through budget management over the planned period of time.
3. The operation or project will not exceed the planned costs if the budget is managed properly. Without a doubt, the man who proceeds to build a house from the ground up without a budget will spend more money than the man who builds the house within a planned budget.

The three variables—the actual *operation* or *project*, the *schedule* or timing of that project or operation, and the *costs*—must be
kept within the parameters of the budget. The budget provides
those parameters; it gives direction. That is why we have budgets.
The mark of good management is reflected in how closely the budgets are followed.

WHEN IS THE BUDGET PREPARED?

Annual (twelve months) budgets may be prepared and finalized
in excess of a year in advance. Biannual (six months) budgets are
usually prepared and finalized mid-period, or three months prior to
the new budget period.

The novice in the budgeting process finds this aspect of fore-
casting, or projecting into the future, the most difficult to come to
grips with. This difficulty is probably due to our natural inclination
to think in terms of the here and now—not nine months downstream or late next year. Experienced, effective managers more often
than not have a reputation of being able to "think ahead" and having certain predictive skills that enable them to anticipate events happening in the future. The average line employee tends to think of
his work in terms of today, whereas the manager thinks of work in
substantially larger blocks of time. Thinking ahead is not necessarily
a measure of intelligence, but rather represents a conditioning and
requisite of managerial responsibilities.

How does one plan for security requirements and costs next year?
How does one plan for criminal attacks and emergencies, which may
or may not occur, which may be large or small in proportion, at times
unknown? The answer is that one does not plan or budget for the
unknown or the unpredictable; one budgets for intelligently anticipated and predictable conditions, based on known conditions in the
present and the past. For example: Security management in the
steel-making industry is planning for the following year. In that year
labor contracts will expire; already issues are surfacing which could
cause serious conflict between labor and management. Under such
conditions, along with a past experience of labor violence, extraordinary security measures should be planned to commence with the
expiration of the contract. As stated previously, *plans must be based*

*on good judgment and good decision-making estimates about the
future.*

As a matter of fact, the predictability of security requirements
for the future is relatively accurate. Truly major emergencies of the
type that would have a serious impact on the budget would be incidents of catastrophic proportion, such as an earthquake or other
natural disaster. Such events, fortunately, are few and far between.

The steel manufacturing security planner, then, can count on
increased security needs when the contract expires; the university
security planner can count on increased problems in June (when the
pressure of final exams is over); the retail security planner can count
on increased needs between Thanksgiving and New Year's. All can
make decisions about plans and costs in the future.

This is not to say that a security department should not have
emergency or contingency plans for major catastrophes. Indeed they
should have. As a rule, however, such emergency plans are broad
and generalized "game plans" which include such things as who
will be in command, reporting responsibilities and channels, specific
asset protection steps and life-saving/first-aid setups. Such a plan
has broad parameters; it is considered a guide or road map and probably has no budget. If the plan must be implemented, costs receive
little attention in view of the magnitude of the problems of property
destruction and the loss of human lives. Dollars, in this context, do
not count . . . at least during the early stages of the emergency.

WHO PARTICIPATES IN THE BUDGETING PROCESS?

There are "bottom-up and top-down" and "top-down and
bottom-up" approaches to budgeting. The latter is preferable because senior management initiates the process by establishing acceptable expenditure guidelines prior to the detailed planning by the
operating or middle management. For a given upcoming budget
year, the general guideline could be that middle management should
continue their cost-effectiveness efforts in all operations as they have
for the previous two years. Following the detailed planning by the
individual managers (Security Manager or Director in our case),
senior management will evaluate and then set the final budget level,
based on the financial outlook for the budget period.

Top-Down and Bottom-Up Process

Phase One: Senior Management—Top-Down
1. Establishes operating guidelines for the Security Department.
2. Establishes acceptable expenditure guidelines; i.e., a given number of dollars.

Phase Two: Security (Middle) Management—Bottom-Up
1. Evaluates the security operation and projects. (*Operation* means a continuing, ongoing function; *project* means a short-term activity; for example, a rape prevention program is a project, not an operation.)
2. Submits courses of action for achieving organizational goals.
3. Costs-out such courses of action.
4. Develops and recommends alternative courses. For example, the initial plan (or course of action) for the rape prevention program came to $700.50. Alternatives include: (a) Do not offer the program at all. (b) Do not buy the film; rent it instead and save $215, bringing the cost to $485.50. (c) Rent the film and do not call in a guest speaker; simply have a rap session following the film, thus reducing the program to $285.50. (d) In addition to (c) above, eliminate posters and announce the program through supervisors, thus reducing the program another $111.50. And so on.

Phase Three: Senior Management—Top-Down
1. Reviews activities, costs and alternatives recommended by security management.
2. Makes decision on the security manager's recommendations.
3. Allocates funds on those decisions, thus establishing the Security Department's next budget.

The entire budgeting process follows a logical or sequential pattern that brings about the interaction between senior and middle management. The sequence is as follows:

1. Planning
   - setting goals and objectives
2. Budget building or budget development
   - evaluating current activities
   - identifying new activities
   - developing alternatives
   - determining costs
3. Evaluation and review of recommendations
   - comparing against original guidelines
   - making decisions re priorities or alternatives
4. Budget establishment
   - allocating funds

The top security executive should work closely with key staff members in the "bottom-up" phase of the process, soliciting input on what the current practices are and what they should be. The executive should be asking such questions as, "Why are we doing it? Why are we doing it *this way*? Do we have to do it? Is there an easier or better way to do it? Can we do it with four men instead of five?" To stimulate the thinking of subordinates in this manner can prove productive in efforts to reduce costs or otherwise effect savings in the function. At the same time, subordinates become involved, at least to some degree, in the budgeting function. Subordinates who participate in budget preparation tend to be more diligent in managing their respective areas of the budget later on.

HOW IS THE BUDGET PREPARED?

Budget costs are classified under one of three categories: salary expenses, sundry expenses, and capital expenses.

Capital expenses will receive little attention in this text because they are usually handled apart from salary and sundry costs. In short, capital expenditures are for physical improvements, physical additions, or major expenditures for hardware. To pay a man for a day's work is a salary expense; to pay for the forms and papers that make up that man's personnel jacket is a sundry expense; and to pay for the metal filing cabinet which houses those personnel jackets is a capital expense. Capital expenses are generally considered "one-time" expenses, whereas salary and sundry are recurring expenses.

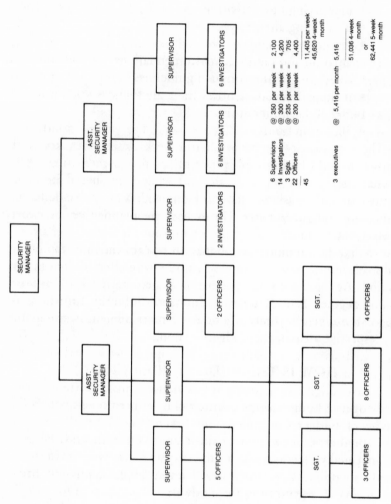

Figure 14-1. Computing salary costs for security department.

Salary Expense Budget

Perhaps the simplest approach to computing salary costs is to count up the security employees, by classification, average out their weekly salary, again by classification, and add it up. (See Figure 14-1.)

Several things should be pointed out with reference to Figure 14-1. First, the computation could reflect anticipated overtime pay *if* there is a history of overtime requirements during the regular pay periods. If that is indeed the case, then the average amount of paid overtime can be included in the weekly totals or could even be averaged out to a monthly total. Thus a four- or five-week month could reflect, for example, $500 in overtime that would be part of the salary calculation for the period. Additionally, if there is any significant amount of overtime or holiday pay due employees during the budget period, that too should be included in the calculations. Otherwise, salary expenses will exceed the planned and approved budget.

A standard requirement is that all budget variances in excess of a predetermined amount must be explained by the executive responsible for the budget, the Security Manager or Director. Failure to calculate Labor Day, Veterans Day, Admission Day, Christmas and New Year's Day in holiday pay due to personnel (assuming those are paid holidays) could result in a significant "overage." It would be embarrassing to explain that such holidays were overlooked during the preparation of the budget.

First rough, then finalized calculations must be transferred to standardized budget forms, controlled by either the company's Controller or Budget Controller. (See Figure 14-2.) Usually one copy (the original) is submitted to the Finance Division and a copy retained in the department. Figure 14-3 reflects that transfer from the drafting stage to the formal stage. Explanations for lines 1 through 10 in Figure 14-3 are as follows:

Line 1— 45 men were used last year and the same number are planned for the coming period. Although the same number of men are being used, the dollar variance represents salary increases due to merit and wage adjustments.

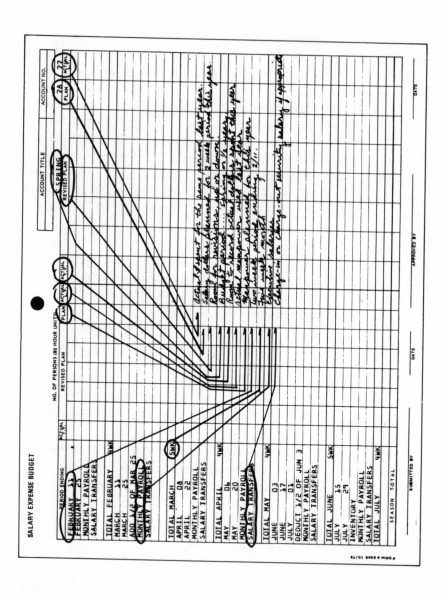

Figure 14-2. Standardized form for calculating salary expense
budget.

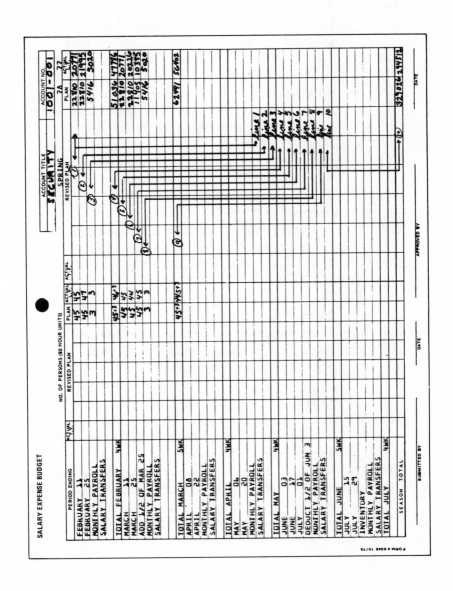

Figure 14-3. Final calculations for salary expense budget.

Line 2— Two extra units (men) were used during a two-week period last year. Because it is only a fluctuation from the norm, we know that it represents overtime.

Line 3— Three security "executives" last year and this year. Again, variance in salary is due to salary increases.

Line 4— Total use of personnel for the month of February, this year against last year, and total salary dollars spent last year against the plan for this year.

Line 5— Same as Line 1.

Line 6— Last year the department was down one man for a two-week period, perhaps because of a personal leave or because an employee quit. Salary reflects his absence.

Line 7— Budgeting on the "4-5-4" plan is most common in business today. This line represents the fifth week in a five-week month, and the odd week is so recorded in dollars.

Line 8— Executive salaries in the private sector are normally treated with confidentiality. Obviously the boss's salary would become common knowledge if recorded alone. In this instance his salary is folded in with that of his two right-hand men. This is a common practice.

Line 9— Same as Line 4.

Line 10— Actual salary dollars spent for the budget period last year, and the planned salary dollars to be spent for the same period this year.

Company-wide pay adjustments or so-called "cost of living" increases, regardless of when they take place, force the revision of the budget at the departmental level. Random increases throughout the ranks, on the other hand, normally do not require a budget revision. Revisions, upward or downward, are recorded on a form designed for just that purpose. (See Figure 14-4.)

Sundry Expense Budgets

All ongoing non-salary expenses are considered sundry expenses. Figure 14-5 reflects a security "supplies" account. Those expenses charged to this particular account are somewhat arbitrary because the division of expenses into given accounts really depends on volume. In a large security organization, it is quite likely that a *separate* sundry account would be established just for uniform replacement and cleaning. In the same organization the "supplies" account would reflect only the first four items listed in Figure 14-5.

In addition to volume, another criterion for creating a sundry account would be the distinctive identity of that account; for example, a "travel expense" account, funds allocated specifically for security executives or personnel to travel between facilities and locations where their presence is required.

Sundry security accounts could include the following:

- Supplies
- Uniforms (replacement or upkeep)
- Travel
- Transportation
    - lease of patrol vehicles
    - maintenance of vehicles
    - insurance of vehicles
- Contract services
    - central station alarm contracts
    - employment screening service
    - polygraph examiner retainer
    - undercover (intelligence) service
- Professional
    - organization membership fees
    - business or professional luncheons and other entertainment costs
    - publication subscriptions.

Again, volume and identity (or, perhaps better stated, those expenses that have a common denominator) dictate the number of sundry accounts a given department might have. If uniform-related expenses are not significant in terms of dollar volume, then those expenses can be budgeted under the next most logical account, such as Supplies (see Figure 14-5). If the only travel expense is an annual trip to a convention, then a separate Travel account would not be justified. The travel expense could be budgeted under Transportation or Professional.

Because budgets are a management tool, it is reasonable to conclude that one very large budget could be cumbersome and difficult

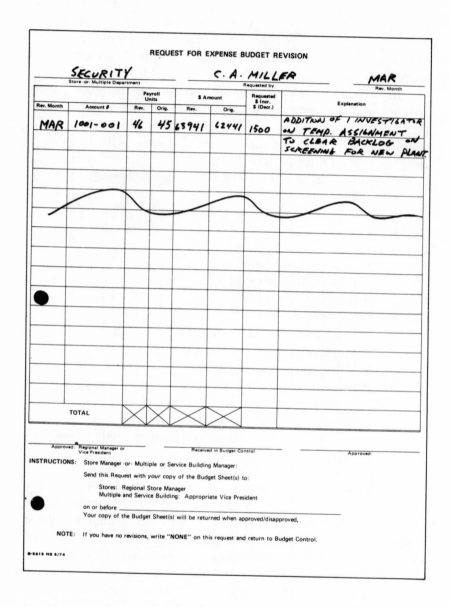

Figure 14-4. Request form for expense budget revision.

SUNDRY EXPENSE BUDGET REVISED PLAN

ACCOUNT TITLE: SECURITY SUPPLIES
ACCOUNT NO.: 1000-50

MONTH	ACTUAL T.Y.	DATE	DATE	DATE	DATE	PLAN	ACTUAL L.Y.
FEB 78						1000	826
MAR 78						1000	1151
APR 78						650	997
MAY 78						950	1084
JUN 78						700	1123
JUL 78						950	1002
TOTAL						5250	6183

BUDGET AMOUNT IN MONTH IN WHICH CHARGE WILL BE MADE — TOTAL SEASON

DESCRIPTION OF ITEM	FEB / AUG	MAR / SEPT	APR / OCT	MAY / NOV	JUNE / DEC	JULY / JAN	BUDGET	LAST YEAR
PRINTING COSTS	200	200	200	200	200	200	1200	1316
DUPLICATING MACHINE RENTAL	150	150	150	150	150	150	900	900
OFFICE SUPPLIES	50	50	50	50	50	50	300	277
TYPEWRITER MAINTENANCE POLICY - ALL MACHINES	50	50	50	50	50	50	300	300
RAILROAD SEAL RE-ORDER	350						350	700
UNIFORM REPLACEMENT		300		300		300	900	111
UNIFORM CLEANING	200	200	200	200	200	200	1200	1371
LOCKS & KEYS		50			50		100	88
TOTAL	1000	1000	650	950	700	950	5250	6183

LIST PRINCIPAL ITEMS INCLUDED IN THIS BUDGET — GROUP SMALL AMOUNTS TOGETHER AND LIST AS MISCELLANEOUS

SUBMITTED BY _____ DATE _____ APPROVED BY _____ DATE _____

8 - 9040 10/73

Figure 14-5. Sundry expense budget for Security Supplies.

to work with, whereas a number of smaller budget accounts are far more manageable and easier to work with. A significant variance in a large "catchall" sundry account would require the manager to track down all sorts of expenditures to find the cause of that variance. If the same variance is in the Transportation account, the tracking time to discover the explanation is reduced. Thus, having a number of manageable sundry accounts is an efficient way to manage money.

As in the salary expense account, sundry expenses should be relatively predictable, based on good planning for the future period as well as experiences of the past. Statements or invoices should never come as a surprise to the account, except for unpredictable emergencies. In those rare cases, the budget should be revised then and there to reflect the increase.

Earlier in this chapter we discussed the three variables in the budget: the operation or project itself, the schedule or timing, and the costs. The timing of costs is most evident in Figure 14–5. So many dollars are budgeted for each month in the upper portion of the form, and then those planned expenses are broken down into specific expenditures, *by month,* in the lower half of the form. If this particular budget is properly managed, then the railroad seals will be purchased in February, not in any other month. And so it is with uniform replacements and the purchase of locks and keys.

A word of explanation is in order regarding the "Actual T.Y. (This Year)" column on the sundry as well as the salary expense budgets. As the manager works with his "tools" month by month, he records his budget management results in the "Actual T.Y." column as the figures become available. Figure 14–6 reflects actual expenses for the first two months of the budget period, as entered by the manager. This obviously serves as a red flag that he has already exceeded his budget by $210, which should force him to look for a comparable savings during the following months so as to come in within the budgeted total amount. Failure to keep a running tab on expenses, sundry or salary, can lull a manager into complacency, and the net results at the end of the budget period can be an unpleasant surprise.

Justifying the Security Budget

It is clear evidence of poor communications between senior

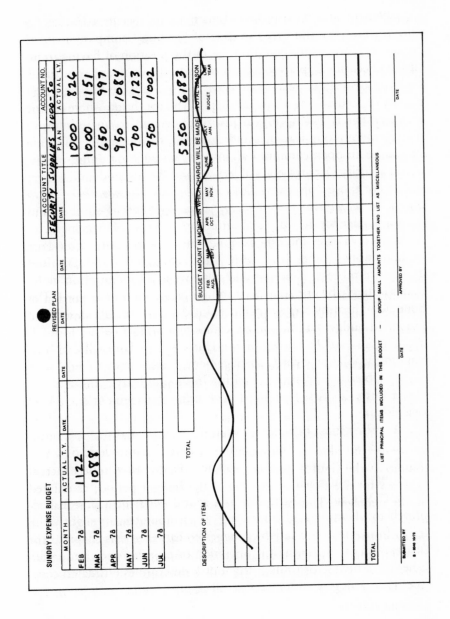

Figure 14-6. Security Supplies budget, showing actual expenses for two-month period.

management and security management if the Security Manager or Director must wrestle over the issue of "selling" company management on security. He should not be in the position of having to justify the Security Department's existence in the company, nor should he have to struggle for his fair share of the available budget dollars to be allocated among all departments. The condition where security management is attempting to justify protection to an uninformed management group is to be avoided; rather, the operating conditions should be an open communications line, an ongoing understanding of the Security Department's objectives and the methods and strategy to achieve those objectives. *The objectives should have been established by management and security, jointly.* This is another illustration of the top-down and bottom-up process. Management initially sets operating and expense guidelines, and Security, after doing its homework, comes back with plans, alternatives and priorities. The entire process, if honest and healthy, is a mutual effort rather than a struggle by one side (Security) for recognition or survival. That honest and healthy process includes, incidentally, the absence of old-fashioned budget padding in anticipation of budget cuts.

Dr. R. Paul McCauley of the University of Louisville, an outstanding scholar of security management, has a theory called "security by Objectives" which is exciting in its simplicity.* In an abbreviated form, his first four steps in security management's approach to a task are:

1. What must be done (or what do we have to do)?
2. How must it be done (or how are we going to do it)?
3. When must it be done (or when are we going to do it)?
4. How much will it cost?

Dr. McCauley's theory forms a sound basis for a practical approach to planning and budgeting. It incorporates, in slightly different form, the three variables suggested earlier in this chapter. The first two questions define the operation or projects; the third establishes the schedule or timing; the fourth determines estimated costs. In essence, this is what a budget is all about.

*McCauley, R. Paul, "Zero-Based Security," *Security World* Magazine, Vol. 14, No. 7, July 1977, pp. 40–41.

Summary

Planning and budgeting go hand-in-hand; a budget is a plan stated in financial terms. Budgeting requires a realistic estimate of programs and their costs, and an allocation of resources to achieve planned objectives.

Because budgets are prepared well in advance, effective budget management requires thinking ahead, anticipating needs based upon relatively predictable conditions. The budget then becomes a tool to insure that plans are carried out. It gives direction to planning by requiring the definition of specific programs, their timing and their costs.

The "top-down and bottom-up" approach to budgeting is recommended. In this process senior management establishes goals and guidelines; the Security Manager provides the detailed planning and cost estimates; senior management reviews these recommendations, establishes priorities, and allocates funds. Where budgeting reflects this interaction of senior and middle management, the protection program will be "presold," based upon a mutual understanding of company goals and departmental objectives.

Budget costs are generally broken down into capital expenses, salary and sundry expenses. The first category of costs are generally fixed or easily determined. The use of detailed records from month to month and year to year makes it possible to arrive at realistic salary and sundry projections. In the area of sundry expenses, such factors as the volume of expenditures in a given category and the distinctive nature of given costs can be used to establish categories. It is generally more efficient to manage a number of smaller sundry accounts than to rely upon large, "catchall" accounts.

Review Questions

1. What are the five elements in the definition of a budget?
2. What are the three variables in a budget?
3. Briefly outline the steps in the "top-down and bottom-up" approach to budgeting.
4. Give an example of each of the following: a capital expense; a salary expense; a sundry expense.

5. Discuss the approach to setting up a salary expense budget. How should overtime expenses be handled?
6. Give four examples of sundry accounts that might be set up for a Security Department.

Chapter 15

Program Management

The Security Department of any company provides a protection program addressing itself to the security needs of the company, based on the presence of known risks. The protection program is comprised of many lesser programs, all somehow interrelated and coordinated to form the entire protective blanket.

Because companies are living, dynamic entities, they constantly change, requiring the security programs to change also. Failure to adjust to the company's constant shifting and movement puts the security efforts out of focus. Example: a firm changes its policy and commences paying its employees on payday with bank checks instead of cash in pay envelopes. The Security Department's "program" for protecting the large cash payroll and everything surrounding that responsibility, including the arrival of the armored car in the morning, filling pay envelopes and disbursement of envelopes, must change.

The need for change in this case may be most evident, but in reality there is a phenomenon of resistance to change. Long after the change is made to payment by check, there is a chance that part of the Payroll Department's program *and* the Security Department's program will still be in place, functioning inefficiently and at an unnecessary cost. Notice that the Payroll Department is included in this program failure. Program failures are not limited to security but occur everywhere in organizational life, particularly in large organizations, with the chance increasing in direct relationship to the increased size of the organization.

With change come new risks, hazards that were not necessarily present prior to the change. The new risk in the example above might well be the vulnerability of the blank payroll checks to misuse. Thus a new security program is needed.

THE INSPECTION PROCESS

Inspection is an important part of the security management process which insures that risks are recognized and covered in security programs, and that programs are necessary and cost-effective. The assessment of what is happening from a security point of view is made through an inspection program. Such an inspection program

1. Must have full support of senior management to bring about necessary change if change is needed.
2. Must be continuous in nature.
3. May be formal or informal.
4. May be structured or unstructured.

Support of Senior Management

If the inspection, which is a close and critical examination or scrutiny, reveals the need for change, then change *must* occur. More often than not the change requires money. Again using the payroll example, safe storage for the blank payroll checks requires a secured storage room or vault. If management fails to provide the necessary funds to construct, modify or otherwise secure a storage area, and refuses to allow the purchase of an adequate vault, the inspection process is compromised. Such an occurrence is not uncommon: a security inspection will reveal a need, but management decides to incur the risk (gamble that the risk factor is not worthy of the expense) and withholds the required dollars.

In the case of the blank payroll checks, to secure or not to secure is the issue. There is no empirical way of determining the extent of the risk; it is judgmental. Management personnel in the finance division could say that the risk is minimal because the checks must be processed through a check-making machine that imprints the amount as well as affixing the indicia-signature, and the machine is always under lock and key. Security's position could be that an employee intent on defrauding the company by means of the blank checks will bide his time until security on the check-making machine becomes lax (and it will). Or Security may point out that check-making machines are available on the market (as well as underground) and the indicia-signature is easy to counterfeit.

Because company management and security management do not concur on the level of risk or the probability of loss does not necessarily mean that company management lacks confidence in its security management; on the contrary. Company management, which is ultimately responsible for the welfare of the organization, is functioning not ill-advisedly or ignorantly, but fully advised by protection professionals of the whole spectrum of risks. Company management makes the final decision as to the dollars they are willing to spend to prevent losses, and the chances of loss they are willing to take by investing those dollars elsewhere in the business. After all, simply being in business is a large risk in itself.

Ideally, security must have the full support of senior management to bring about change if change is needed. The "if needed" should be the only opposing issue between security and management. The good security executive more often than not will obtain management's agreement that the change he recommends is, in fact, needed.

Continuous Inspections

To keep pace with the constant changes in the organization, changes in our society in terms of attitudes, life-styles and moral values, and the rapidly advancing technological modifications all around us, the inspection process must be an ongoing, never-ending activity. The larger the organization, the more reasonable that statement sounds; the inspection task never seems to be finished. For smaller organizations, however, down to the one-man operation, the continuous inspection process appears less reasonable. Yet even the smallest security department has a host of internal (security department) and external (company) areas to inspect—and inspect in great detail, too. The added advantage in the smaller firm is that the security inspection may also serve the purpose of an internal audit, which is usually conducted in a large company by internal auditors from a sub-unit within the finance division.

Inspection is one area where it is fair to say that one's work is never done.

Formal or Informal Inspections

A *formal* inspection is one to which some fanfare is attached; it is usually preceded by an announcement, and the unit under inspection "prepares" for the event, including some extra housekeeping activities that would not otherwise happen at that point in time. To add to the importance of the occasion, a senior executive may accompany the security executive on the inspection, thereby encouraging total cooperation on the part of the unit management.

For the company which has just upgraded the security function, hired a new security administrator, or initiated a security department and program with a new security chief, the formal approach to inspections is most desirable, primarily because it tells the company how senior management feels about protection, and thus establishes the desirable climate.

Informal inspections are usually the result of a long and firmly entrenched inspection program, understood by all and accepted as part of the organizational life. The inspections have been stripped of all the external trappings of importance, but their functional importance has not been lessened in any manner. They are seriously and quietly executed in a spirit of understanding and cooperation.

Structured or Unstructured Inspections

A *structured* inspection, as opposed to an unstructured inspection, is one that moves systematically, perhaps even rigidly, from one designated inspection area to the next and from one inspection point to the next. The following could be part of such a structured inspection:

Warehouse Exterior
1. Fencing
    a. vegetation growth
    b. general conditions

     c. additions or deletions?
     d. evidence of penetration?
  2. Gates
     a. gate schedules
     b. inventory locks
     c. lock schedules
     d. key controls
     e. gate assignments
     f. gate traffic logs
       1) train
       2) truck
       3) personnel

Two examples of checklists for use in a structured inspection are included as appendixes to this text. Appendix B is a more comprehensive example of a structured inspection or security survey. Appendix C is another structured "checklist" approach to inspecting a retail operation.

The *unstructured* inspection, in contrast, would approach the warehouse unit in a more random manner, with less methodical attention to small specifics. The experienced eye of a top security professional would note at a glance, without following a checklist, that weeds and other vegetation against the fence needed clearing.

It stands to reason, then, that the decision as to which type of inspection format is needed depends a great deal on the expertise of the security executive who is involved.

Who Conducts the Inspection?

Ideally the Director or Manager should conduct the inspection, along with, *in every case possible,* the next-ranking man in his organization. For example, in a very small department, with a chief and a uniformed staff of six men, including a sergeant, the sergeant should accompany his chief.

Why should the manager himself conduct the inspection? It would seem that a subordinate could easily follow the structured inspection and its checklist.

Certainly any number of security officials, down to supervisors, can conduct inspections. But the lower in the ranks the function is delegated, the less important the event becomes in the eyes of the management area under inspection. This is one reason why the head of security should conduct inspections. A second reason is that company management looks to its security manager for *his* expertise and wisdom when it comes to protecting the company. His involvement in assessing risks and countermeasures assures them of the *best* assessment.

Why have a second man along on the inspection? There are three reasons. First, the experience itself is an outstanding training activity. After accompanying the manager on a number of inspections, the second in command—the lieutenant, for example—could move into that function easily and confidently. Secondly, besides gaining valuable information about the entire process of risk assessment and program evaluation, the lieutenant gains stature in the organization because of his relationship with the manager, who in-variably holds great respect by virtue of responsibility and position in the company. Finally—and this is particularly true in small security organizations where the next in command is not necessarily the heir apparent—he becomes increasingly "sensitized" to conditions he never recognized before and conditions that never occurred to him. This reason, like the first, is a form of training. In the former example, however, the purpose of the training was to prepare the lieutenant to move up to the top; here the purpose is more to increase the level of awareness of a line-type supervisor and improve his efficiency on the job.

Another dimension can be added to the inspection process by having appropriate security supervisors quietly make an inspection in advance of the real one, using the structured format as a guideline, and letting them subsequently compare their results with the manager's. The manager had better be thorough, however, because if he overlooks areas which the supervisors found to require change or correction, his failure tends to discredit his ability in the eyes of those subordinates.

In a large organization in which the warehouse is to be inspected, the following would probably comprise the inspection party: Security Manager, Assistant Security Manager, Security Supervisor whose area includes the warehouse, the warehouse superintendent and his assistant.

In a small company the inspection party might include the

Security Manager, his Sergeant, the warehouse foreman and his superior, who probably has numerous responsibilities which include the warehouse.

In some situations, particularly in the establishment of new programs or a comprehensive re-evaluation of an existing program, it may be necessary or advisable to engage the services of an outside security consultant to conduct a security survey. In such cases the security manager or director will, of course, work closely with the security consultant and with company management.

The entire inspection program, again, has as its objectives (a) the assessment of risks, and (b) the assessment of their counter-measures, usually security programs. We will examine both aspects of this assessment.

ASSESSMENT OF RISKS AND COUNTERMEASURES

Risk Assessment

Inspection reveals conditions brought about by any number of things, such as the company's decision to pay by check instead of pay envelopes, which may pose a security risk. The possibility and probability of the risk resulting in a loss, and the magnitude of the loss, depend on the risk itself.

For example, every security executive would agree that finding a cigar box used in the purchasing department as a repository for a

The risk	Is it possible to have loss?	Probable?	What would be probable loss?	How much to countermeasure?	Cure risk?
Open petty cash fund	Yes	Yes	$15	$4.50	Yes
Blank payroll checks	Yes	Yes	Many thousands	$385	Yes
Unprotected skylights in grain warehouse	Yes	No	Under $3,000	$5,000	No

petty cash fund represents a risk. If the fund amounts to $15, the loss would be $15. That is clear. And to "cure" that risk or provide a countermeasure (perhaps in the form of a lockable tin box with the keys in the possession of the accounts payable manager) is a relatively easy and inexpensive thing to do. In the case of the blank payroll checks, on the other hand, the possibility of loss, the probability of loss, the potential of loss, and the cost to cure the risk form a more difficult equation.

It is not within the scope of this text or chapter to delve deeply into the exacting science of risk management, a subject in itself. The key point to be made is that from the discovery of what appears to be a risk, to the decision as to what action to take (i.e., cure it, minimize it, or live with it), there is a close interaction between security management and company management. This interaction takes place in the office of the security head's superior, where the final decision is made, particularly if the countermeasures involve capital or expense funds.

Selection of Countermeasures

There are four possible cures or countermeasures for every risk: procedural controls; hardware (fences, gates, locks, keys, etc.); electronic systems (alarms, access controls, CCTV, etc.); and people.

In the case of the petty cash fund, the countermeasure was hardware, in the form of a lockable tin box. In the case of the payroll checks, not only would hardware (vault or vault-type room) be a countermeasure, but so would a control procedure such as the accounting for checks by check serial numbers, withdrawal of checks from the vault by number (by batch) and the signing for such withdrawal by the appropriate employee, a prompt check reconciliation program after return from the bank, etc. The unprotected skylights might require hardware (an inside latch or an outside lock or padlock) and/or electronics (an alarm).

One of the important ongoing responsibilities of the Security Director or Manager will be to evaluate, select and recommend appropriate deterrents for each significant risk from these four categories of countermeasures.

Procedural Controls. Policy tells us *what* we must do, whereas a procedure tells us *how* we are going to do it. Procedural controls are intended to define how any activity is to be carried out in such a way as to prevent or expose any violation of policy (and attendant potential for loss).

For example, a gambling casino's policy is to insure that change-girls are accountable for their individual "bank." To make them accountable, which in and of itself contributes to the prevention of theft, they must follow a structured procedure—a procedural control. This procedure might spell out the specific amount of the change-girl's fund or "bank." It would then require that she check out her bank from the head cashier upon reporting for work, count it to insure the full amount is there, and sign a designated form that she did receive the fund and the amount recorded was therein. Such a form should also be countersigned by the head cashier and placed in a certain part of the vault, to be removed only when the fund is returned at the end of the shift, at which time the fund is then counted by the head cashier, signed off by same, and countersigned by the change girl. The procedure might additionally require that the form be filed and retained for ninety days. Each of these steps is designed to establish accountability for valuable assets (in this case, cash) and verification of activity by more than one source.

Since it is often just as easy and involves little or no additional expense to establish a controlled procedure as an uncontrolled one, this type of countermeasure is generally the least expensive. (It should be noted that procedural controls in some situations may be automated, as when a computer is programmed to control and monitor the issuance of purchase orders, shipping invoices, bills of lading, receipts and other paperwork involved in a transaction. In this circumstance the procedural controls have become *electronic,* moving into a higher category of expense.)

Hardware. Many loss risks can be significantly reduced by the relatively simple application of some form of hardware, from a padlock on the company's gasoline pump to a perimeter fence with adequate lighting. Hardware is most common in the average person's defensive strategy in his private life. Lockable suitcases, chains and locks to protect bicycles, bars or decorative screens on residence windows, front door peepholes, night latches, outdoor lighting—all of these constitute hardware. In the business environment such physical protective measures may become more sophisticated, progressing to security containers (safes, lockable file cabinets and vaults) and other equipment or devices. Still, hardware is the second least expensive among the four basic countermeasures.

In modern applications security hardware is often combined with electronics; for example, a CCTV-monitored truck gate that is electronically (and remotely) controlled, or a fence whose protection as a physical barrier is supplemented by an intrusion alarm.

Electronics. In addition to closed circuit television, electronic countermeasures include such devices as automated access control systems and the whole spectrum of alarms. These include intrusion alarms, motion detection alarms, sound or vibration detection alarms, smoke detectors, heat detectors, waterflow alarms and others. Electronic devices constitute the fastest growing category of security countermeasures. While more expensive both to install and to maintain than procedural controls or basic hardware, they have become a fact of life for almost all businesses, large or small.

Alarms in particular were originally designed to replace the people who were previously deployed and utilized to provide precisely the same kind of "alarm" or warning service. Electronic alarms have proved to be better than people for a number of reasons. They are less costly (when compared to the annual salaries of personnel required to perform the same function); they will not fall asleep; they are always on the job despite deep snow, slippery streets or a death in the family; and they are honest.

Personnel. Ironically, the utilization of people as a security countermeasure can be the most efficient and effective strategy or, depending upon the circumstances, the poorest.

Because of the ongoing expense of personnel (not only for salaries but also for fringe benefits, vacation and sick leave, supervision and replacement), every effort should be exercised to cure risks whenever possible by every means other than by utilizing people. The rule of thumb is: use people only in those areas where procedural controls, hardware or electronics cannot be employed more efficiently.

Obviously, there are security functions for which people are the best and sometimes the only countermeasure. The greatest attribute of people, one which can never be replaced, is their ability to exercise *judgment.* In that capability lies a critical factor in the decision to use people. Wherever judgment is essential in carrying out a security function, people should be utilized. A common example might be the job of overseeing employees as they leave work in a production plant, in terms of inspecting lunch pails and other containers. Personnel are essential for a variety of other roles which cannot be effected by procedures, hardware or electronics. Among these functions are guard posts and patrols, inspections, investigations, prevention of criminal attacks, maintaining order, crowd control, etc.

Assessment of Countermeasures

The other side of the inspection coin is the examination of existing countermeasures, usually protection programs and activities, originally set into motion to cure the known risks. Whereas the discovery of risks usually comes from conditions that are observable or comprehensible by virtue of what has happened, what is happening and what could happen, countermeasures are best assessed through an analysis of the countermeasure activity itself. This analysis is usually accomplished by asking questions. *The primary and most devastating question is "why?"* Every countermeasure, every security program, should be subjected to the following questions:

Why are we doing it?

Must we do it at all?

If we must, is there a better way?

Is there a less expensive way?

In one organization, for example, the new Security Director asked the supervising agent of the main complex what the guard on the back side of the complex was doing there. The Director was told it was a necessary post because the guard logged all trucks coming through that gate, opened the gate in the morning, locked the gate later in the morning, logged trucks coming from the main entrance that serviced the far side of the complex, and opened the gate late in the afternoon to accommodate employee vehicles. Additionally, the guard controlled traffic when the railroad brought in freight cars and took out the emptied ones. Besides, the Director was told, "We've always had a man on that post. That's why we built him a guard shack and installed a telephone."

By holding this particular assignment up to the four questions above, it was discovered that the guard was a holdover from a former

procedure and that his work, such as logging trucks, was a total waste.
No one ever checked his logs, and they contained no more information than that gathered at another location. The roving patrol officer
could open and close the gates according to schedule, and the railroad's brakeman stopped what little traffic there was when the box
cars were being moved. Thus the company eliminated one guard, as
well as the expense of one telephone, and was able to tear down one
guard house that consumed energy needlessly with its electrical
service.

Elimination of functions is obviously the ultimate in reducing
costs, but the opportunity for such action proves to be rare. By per-
sistent application of the *why* test, however, limited opportunities
will constantly present themselves. Another outside guard operation
was modified, for example, by replacing weekend guard coverage
with an electrically controlled gate, CCTV and communication
phone. The guards had been poorly utilized during this slow period
and were not only unproductive but unhappy with that assignment
because it was boring. These hours were eliminated from the salary
budget. The capital expense for the hardware amounted to less than
the annual cost of one 40-hour man unit. Thus in one year the
equipment paid for itself and thereafter effected a cost savings to the
company.

There is absolutely nothing within security's spectrum of programs that should be immune from this inspection. The difficulty is
not so much in the application of the inspection as it is in getting
managers and supervisors at every level in the security pyramid to ask
themselves and their respective subordinates, "Why are we doing it?
Must we do it at all? If we must, is there a better way? Is there a less
expensive way?"

Interestingly enough, whereas managers and supervisors for
some reason tend not to question their pyramids (and programs
therein), line security people do see better ways. The effective security
manager will encourage employees to come to him with their ideas.
Too often they have already suggested improvements that were
ignored or rejected by their supervisor; to make the suggestions to the
manager might be misconstrued by the line supervisor as insubordi-
nation. If an open climate of communications is established, however, as discussed in Chapter 12, there will be a constant flow of new
ideas.

Inspecting for Compliance with Procedures

Whereas our discussion of inspections up to this point has focused on finding new risks, primarily brought about by change within the organization, there is another very important dimension to the inspection process. Inspection provides the additional benefit of determining *compliance* with already existing countermeasures which are known to be sound. This type of inspection is executed at the line or first line supervisory level.

We may be satisfied, for example, that the locks on the gates are proper, and the control of keys is properly spelled out, and the gates are scheduled to be locked and unlocked at the right time—but *are these things happening as they should?* As we observed in an earlier chapter, "people do not do what you expect, they do what you inspect."

The Security Director may set up a procedural control that requires the payroll department to secure the blank checks behind locked doors at the end of the day. But do they? The Security Director's inspection of the payroll area does not insure compliance later on. But inspection by his people will tell him whether the payroll department is ignoring or forgetting the procedure or in some other way still contributing to the original risk. How security management responds (or reacts) to such information is absolutely vital, for two reasons:

1. If security management fails to act on the information that the procedure is not being followed, they permit the risk to exist, which is counterproductive and inexcusable.
2. The line people who make the discovery of noncompliance and consequent risk, and see their management's failure to take prompt corrective action, become discouraged and say "what's the use?" The downstream result is that the line employees lose interest and risks increase everywhere.

If security management is not going to follow up on inspection deficiencies, then it should not ask line employees to make inspections. Further, compliance inspections come to the attention of employees who work in the area being inspected, and if they come to realize nothing is going to happen because of compliance failures, they will laugh at the inspection as nothing more than an exercise.

If a condition or procedure is worthy of being inspected, it is

	DE TM	DE YTD	DE LYTD	MIS DET TM	MIS DET YTD	MIS DET LYTD	TOT DET TM	UC DET YTD	UC DET LYTD
N-1011	1	4	2	0	0	0	4	0	0
N-1012	1	4	2	0	0	0	1	0	0
N-1013	3	8	3	1	3	3	9	1	0
N-1014	1	4	0	0	4	2	2	0	0
N-1015	0	9	6	0	2	0	1	0	2
N-1016	6	41	25	0	3	5	7	7	4
N-1017	1	27	0	0	1	0	8	4	0
N-1018	0	0	1	0	0	0	1	0	0
N-1019	0	2	3	0	0	0	3	0	0
N-1020	0	1	5	0	3	0	3	0	0
N-1021	0	2	0	0	1	0	0	0	0
NORTH DISTRICT TOTALS	13	102	47	1	17	10	39	12	6
S-1111	1	5	2	0	0	3	4	0	0
S-1112	0	5	1	0	0	0	0	0	0
S-1113	0	4	2	0	0	0	2	0	0
S-1114	1	4	2	0	0	0	1	0	0
S-1115	1	8	7	0	0	0	2	1	2
S-1116	0	2	1	0	0	0	0	0	0
S-1117	0	1	3	0	0	0	2	0	0
S-1118	2	3	6	0	0	0	3	0	2
S-1119	0	0	0	0	0	0	0	0	0
SOUTH DISTRICT TOTALS	5	32	24	0	0	3	14	1	4
C-1211	0	10	12	3	4	3	3	1	0
C-1212	1	4	7	1	2	1	2	0	0
C-1213	2	4	3	0	0	1	2	0	0
C-1214	0	1	2	0	0	0	0	0	0
C-1215	0	2	1	0	2	0	0	1	0
C-1216	2	12	19	0	0	5	2	1	0
C-1217	0	5	5	0	0	0	0	1	2
C-1218	0	2	1	2	2	0	2	0	0
C-1219	0	0	0	0	0	0	0	0	0

	DE TM	DE YTD	DE LYTD	MIS DET TM	MIS DET YTD	MIS DET LYTD	TOT DET TM	UC DET YTD	UC DET LYTD
CENTRAL DISTRICT TOTALS	5	40	50	6	10	10	11	4	2
E-1311	0	5	12	1	1	6	5	0	0
E-1312	0	5	10	0	2	0	0	0	0
E-1313	0	6	1	1	2	1	3	0	0
E-1314	0	1	1	0	0	0	0	0	0
E-1315	0	0	0	0	1	0	1	0	0
E-1316	0	0	0	2	5	0	3	0	0
E-1317	0	14	1	2	5	0	3	0	0
E-1318	5	13	4	0	0	0	7	0	0
EAST DISTRICT TOTALS	5	44	29	6	16	7	22	0	0
W-1411	4	26	13	0	0	3	9	4	0
W-1412	0	21	9	0	0	4	8	2	0
W-1413	1	15	5	0	0	2	4	3	1
W-1414	2	2	2	0	0	0	2	0	0
W-1415	0	4	4	0	0	0	3	0	0
W-1416	0	4	7	0	0	0	0	0	0
W-1417	1	5	6	0	0	0	1	2	0
W-1418	1	4	5	0	0	0	1	0	1
W-1419	0	14	12	0	5	0	0	0	9
W-1420	4	8	0	0	0	0	7	2	0
WEST DISTRICT TOTALS	13	103	63	0	5	9	35	13	11
GRAND TOTAL	41	321	213	13	48	39	121	30	23

Figure 15-1. Statistical record of detection activity in different units of a large organization.

worthy of a prompt follow-up by management. Many security programs, always aimed at reducing risks, often make non-security people's work more difficult—requiring them to keep certain doors locked, for example. It is inconvenient for them to lock and unlock a given door, so human nature steps in and they leave it unlocked. Security finds the unlocked condition in its inspection and reports it. Prompt action corrects the offender, improves security by reducing the original risk, and brings a degree of respect to the inspector. If the line employee believes, based on fact or fiction, that his inspection report will not be acted on, he will ignore the very security risks he is there to detect.

STATISTICS IN PROGRAM MANAGEMENT

Statistics constitute another tool in managing security programs. To be an effective and meaningful tool the statistics must be designed to reflect what a given program is or is not doing—month by month, this year compared to last year by month, and cumulative year to date. If improperly designed, the quantitative figures will either be meaningless or deceptive. Once the statistical format is in place, it too should be inspected on an ongoing basis to insure it has not outlived its original purpose, that it accurately reflects current activities and programs.

Not only must the statistical presentation reflect desired and necessary information and currently reflect activities of an in-place program, but it must be *used*, otherwise the value of the statistics is lost.

Figure 15-1 is an important management tool in determining how effective the detection program is in one large security organization located in many cities, divided into districts. Reflected in the chart are the number of dishonest employees (DE) apprehended this month (TM), year to date (YTD), against last year to date (LYTD), miscellaneous detections (MIS DET) this month, year to date against last year to date, total detections (TOT DET) this month, and detections made by undercover agents (UC DET) year to date against last year to date.

In designing this statistical form, security management was concerned not only with totals, but specifically with detection activity in every single unit by district, by time (month, year to date, and last year to date), by employee and by other (misc.). It was also interested in one sub-classification or technique—that is, apprehensions of dishonest employees by the use of undercover agents.

The undercover data are an excellent example of designing statistics to serve a purpose. In this case the statistics do demonstrate the effectiveness of that particular technique when compared to the budget dollars spent for undercover agents. The last two columns alone tell us a great deal; i.e., the undercover program has improved on the whole, as reflected by the increase in detections for the same number of budget dollars (not reflected in the chart but known to management). They also reflect a problem in the East District and a lesser problem in the South District. What is the problem? Is it poor security supervision in those districts? Are personnel managers in those districts disclosing the identity of the undercover agents, thus defeating the program? It is improper tabulation of statistical data? Whatever the cause, the statistics have waved a red flag that can lead to discovering and solving the problem.

These statistics also reflect another interesting story behind the sharp increase in apprehensions of dishonest employees. The Security Department pulled together, at an increase in the salary budget, a special detection unit of highly skilled security people to form an elite squad. Their impact on the department's overall effectiveness in detecting internal dishonesty is clearly seen in these statistics. Not only can such statistical tools tell the administrator his programs are working, but they also tend to serve as cost justifications.

Summary

As organizations change, so do security needs. *Inspection* is the ongoing process which ensures that new risks are recognized and that established deterrents remain necessary and cost-effective.

To keep pace with organizational change, inspections must be continuous. The inspection process must have the full support of company management and the active participation of security management. Such inspections may be *formal* or *informal*, *structured* or *unstructured*. The structured inspection moves systematically from one area of exposure to another, following a detailed checklist.

Risk assessment evaluates the probability and cost of potential loss. From this evaluation comes a decision to adopt deterrents.

Countermeasures may involve procedural controls, hardware, electronics or (the most costly) security personnel.

The inspection process also includes *assessment of existing countermeasures*. Every security program or practice must be subjected to the basic challenge: Why are we doing this?

Inspection also verifies *compliance* with protection programs. Are they being carried out as planned? Such verification comes at the supervisory and line employee level as well as from management; the effective manager will be responsive to this input.

Statistics offer another tool for ongoing evaluation of protection programs. Statistical information, too, must be subject to inspection to ensure that it is up-to-date, and it must be *used* if it is not to be a meaningless exercise.

Review Questions

1. Define a structured inspection as contrasted with an unstructured inspection.
2. Ideally, who should conduct the inspection? Why?
3. What are the four types of countermeasures? Give an example of each type. How do the costs of each compare?
4. Discuss the possible consequences of security management's failure to follow up on inspection deficiencies.
5. Discuss the role of statistical tools in program management.

Chapter 16

Investigations

The investigative function is, of necessity, different in each type of industry or business within the private sector, as well as different in various companies in the same business. The scope of the Security Department's investigative mission is unique to each company. The composition of the investigation unit of a credit card company's security organization, for example, would be significantly different from the investigation unit in a major amusement park complex. Each would be structured to meet organizational objectives. Obviously, the primary investigative work in a credit card company would be forgery-oriented, while the primary effort in the amusement park environment might well be internal dishonesty. And so it goes across the broad spectrum of the private sector.

Depending upon management's perception or knowledge of the Security Department's investigative capabilities, such skills will invariably be drawn upon for what one might classify as secondary missions, or assignments that do not necessarily help achieve organizational objectives. An interesting example was a recent case where a company's chief executive officer asked the Security Manager to look into the burglary of the executive's apartment. Not surprisingly, the city police had neither the time nor the manpower to give the burglary any special attention. Standard procedure in the city in question was to have a uniformed officer in a patrol unit respond to the scene and take a crime report from the victim.

The security investigator proceeded to the scene and rang the next-door neighbor's doorbell. The neighbor provided some startling information. She had seen the burglars carrying items out of the building into a waiting truck and had telephoned the local police station. The busy desk officer had asked the neighbor how she knew in fact it was a burglary, and she replied that she did not know "in fact." The officer dismissed the incident as being merely a well-intentioned neighbor with an overactive imagination.

Needless to say, when the neighbor furnished the security investigator the out-of-state license number of the truck used in the burglary, the investigative efforts were assured success. Shortly thereafter, the chief executive's property was recovered and the suspects apprehended. Management's previous confidence in the Security Department's investigative capabilities was dramatically reaffirmed.

The police could have done the same thing the security investigator did—but they did not, for a variety of legitimate reasons, including manpower resources. The failure of the municipal police to cope with a variety of other crimes against property, some of which directly affect a company's organizational objectives, gives rise to the need for investigative talent that takes the time to look into criminal conduct affecting business. This kind of investigative talent must come from the private sector's security organizations.

There are seven general investigative categories in the private sector:

- Applicant background investigations
- Internal criminal attacks
- External criminal attacks
- Corporate integrity investigations
- Corporate liability investigations
- Labor matters
- Miscellaneous investigations.

Applicant Background Investigations

One of the most important loss prevention activities a company can engage in is the thorough screening of its most valuable resource, its employees. Companies have varying types of background investigation programs, depending upon company size, turnover rate, nature of business, government requirements, and the sensitivity of positions being filled. The critical importance of this activity is reinforced daily from company to company across the country. Imagine, for example, the reaction of management in a major retail firm when advised by the Security Department that background investigation has disclosed that the new employee in the Children's Shoes department has a criminal record of child molestation!

Obviously, then, the objective of applicant background investigations, also known as screening, is to determine the suitability of applicants for employment or promotion. There are two aspects of suitability: the apparent and the real. The apparent is what the employment interviewer perceives through examination of the applicant's documented statement of his background (the employment application and resume) and the interviewer's assessment of the candidate during an employment interview. If the candidate does not possess apparent suitability, his candidacy will cease. If he does possess apparent suitability, the next step is the background investigation. The background investigation delves into those factors that support or fail to support real suitability, seeking answers to the following questions:

1. Is the applicant really who he says he is?
2. Does the applicant really have the work experience he claims he has had? For the length of time he claims?
3. Does he have the education he claims?
4. Does he possess the skills he claims?
5. Is he financially stable or does he have a history of credit problems?
6. Is his apparent good character genuine or does he have a criminal history? What is his reputation in his neighborhood?

Once these answers are found, the results are forwarded to the party who makes the employment or promotion decision. Security should not act as the trial court or the board that decides who is to be hired. Security is the fact-finding body. It validates or invalidates the applicant's apparent suitability.

Internal Criminal Attacks

Unfortunately, no company is free from the threat of employee dishonesty. Employees at all levels of the organizational structure can and do succumb to the temptation to steal. Obviously, some businesses by their very nature have greater exposure to internal dishonesty and thus direct more effort to the prevention and detection of such activity. Irrespective of one's perception of vulnerability, however, *every business organization*—including the processor and packager of manure—is subject to internal criminal attacks. Objects of attack go beyond a company's product, merchandise, negotiables or currency: tools, equipment, materials and supplies, as well as personal effects, are all targets for dishonesty.

Internal criminal attacks are not limited to theft. Sabotage, various sex offenses (such as obscene notes sent through inter-office mail), and falsification of documents, constitute but a small sampling of internal attacks that require professional investigative talent which the public sector simply cannot provide.

External Criminal Attacks

Every business establishment is the potential target of the community's criminal element, whether it be a simple sneak-thief looking for employees' pocketbooks or other personal belongings, or the professional burglar who goes about his business with well-laid plans. Certainly there is local police involvement when a company is the victim of a major burglary or robbery, yet it is revealing to observe how much further the Security Department's own investigation commonly goes in terms of details, and how the gathering of those details frequently makes a significant difference in the case.

Whereas the objective of the police is to identify and apprehend the perpetrator of the crime, the security investigator has a second and equally important objective: to identify the weakness or vulnerability that contributed to the success of the criminal attack, and, once the weakness is identified, to make appropriate recommendations to management to correct the vulnerability. Such recommendations could be as simple as expanding the radius of motion detection hardware in the facility's alarm configuration, or as sophisticated as suggesting that company negotiables include engraved patterns in the document's design as a defense against recurrence of counterfeiting.

Corporate Integrity Investigations

This category could also be called Internal Misconduct Investigations, because it deals with investigations that focus, again, on employee behavior. In this case, however, the behavior is not strictly criminal in nature. Rather, it is behavior that is organizationally unacceptable or counterproductive. And more frequently than not, the subject employee is at the managerial or executive level.

The actions of employees who, by virtue of their position, have a very visible profile reflect on the business itself. If the chief executive officer wants the actions of a member of management investigated, it is best if possible to keep the matter "in-house." Such investigations are exceedingly sensitive and great discretion must be exercised.

Other investigations in this category include cases of conflict of interest and unauthorized release or misuse of proprietary information.

Corporate Liability Investigations

In this category we are concerned with the company's *civil* liability. Skyrocketing insurance premiums, with the resultant larger deductibles, mean that a company is virtually self-insured. It is not uncommon today for a major retailer to carry a false arrest policy with a $100,000 deductible. The weight of insurance-type investigations shifts, then, from the insurance company to the insured.

Exposure to civil liability covers a broad spectrum indeed: from "slips and falls" on company premises to company vehicle accidents, from false imprisonment to libel, from a patent infringement action to a battery suit, all requiring the collection of information. That collection process, the investigation, may be accomplished by the insurance carrier's investigators, it may be a joint investigative effort, or, as suggested above, it may fall entirely on the company insured.

Labor Matters

Whether or not a company is organized, the investigation of labor matters is an extremely sensitive issue due to numerous implications which may be raised under state or federal labor laws. Consequently, such investigations would normally be conducted at the direction of or under the guidance of the company's Labor Relations Department and/or company counsel.

Where disruptive events occur during a strike or an organizing effort, the investigative arm of the company is the best equipped and prepared to document the occurrence and collect whatever evidence there may be, such as photographs and signed statements, which subsequently could be presented to the court for possible injunctive relief.

Miscellaneous Investigations

There is a relationship between senior management's awareness of and confidence in the Security Department's investigative capabilities and the quality and quantity of unusual requests and demands made on that capability.

One example is a case referred to the investigative unit of a large Eastern corporation. The investigative unit was housed some 50 miles from the corporate offices. This unit had a highly professional staff and had earned a reputation in the corporation of being able to do "almost anything." One morning a call was received at 10:30 from the corporate president, advising that he was en route to a midtown New York office building to engage in negotiations to purchase that building. His corporate real estate staff had failed to determine the size of the building's freight elevators, and such information was vital. The president not only needed the specifications, he did not want it known that the capacity of the elevators was of any particular concern to him. His call to the investigative unit was a request for them to secure the needed information and put it in his hands as he arrived at the building at 11:30 a.m.

The building in question was more than an hour's distance from the investigative unit's office. During a hastily called investigative strategy session, someone mentioned that New York City had a very strict elevator inspection code and that the city's Building Department would have a record not only of the elevators of every building, but of their size, capacity and date of last inspection.

The requested information was in the hands of the president within 15 minutes, and the reputation of the unit's ability to do "almost anything" was perpetuated.

This case points out that, with a logical approach and intelligent use of available resources, almost any problem can be solved. That is what investigation is all about: problem-solving. It also demonstrates that once the reputation of sharp investigative skills is established, these skills will be tapped continuously by the organization for a wide variety of challenging reasons.

Establishing the Investigative Function

As stated earlier, the character of the investigative function is unique to each company. Due to its uniqueness, the scope of the function must be clearly defined by management, as must the specific reporting channel through which the investigative reports are to be distributed.

Another early decision to be made is: should the company attempt to retain its own investigator, or should investigations be contracted out to agencies which specialize in providing such services? More often than not, companies do both. It will select those types of investigations it wants to perform "in-house" and those which it will farm out. Many Security Managers have available to them "freelance" or individual investigators who are available for special assignments which otherwise would drain all the time of an in-house investigator, who must handle a constant workload of bread-and-butter cases.

In-house investigators are preferable to contract investigations, for three reasons. First is the economics of the matter. To contract cases out to a commercial firm on a continuing basis often becomes cost-prohibitive. Those same contract dollars could be turned into salary dollars, with a savings.

Second, it is an advantage to have a company investigator who is knowledgeable about the company, its products, service or objective, the internal paperwork flow, policies and procedures. An outside contract investigator would face precisely the same obstacles as a police detective in tracking computer data backwards through keypunch to audit department, for example.

Third, the company investigator is a known quantity. His employment history, education, limitations (as such), biases, intuitiveness—all these factors end up in his final product, the completed investigation. The Director or Manager can then pick up or embrace his findings and carry them to senior management. In a very real sense, the investigator's work becomes the manager's work, and the manager can better identify with and support the investigator's work if he is a known quantity.

Undercover Investigations

The advantages of the company investigator do not always hold true in some types of undercover investigation where it is necessary to "place" an investigator in the role of an employee. Particularly in the smaller community or company, it may be difficult if not impossible for in-house investigators to maintain a plausible "cover" for long. The contract agency can assign investigators in an undercover role who are completely unknown to any company employees.

The contract undercover investigator in the private sector might more accurately be called an *intelligence* agent. His function differs from that of undercover investigators in the public sector—usually narcotics or vice officers—who gather evidence for their own cases, then follow through on the cases from arrest to prosecution. Contract undercover agents usually supply information to the company's security investigators or to company management. It is the company, not the undercover investigator, which decides how the intelligence gathered is to be used, including when or if an arrest is to be made.

Confidentiality of Investigations

Since most investigations are either conducted at the direction of management or are reported to management for its decision or action, the information developed during the course of the investigation must be treated as confidential. Few things can injure the investigative function's reputation and standing within the company more quickly or completely than to have the results of the investigation known generally throughout the company before management receives the official investigative report.

It is normally management's prerogative to decide when such information is to be released, to whom it will be released, how it will be released, or whether it is to be released at all. If information gathered in a routine investigation is leaked, then what degree of confidence can senior management have in the Security Department's reliability in handling information deemed highly sensitive in nature?

A good rule to follow is to treat *all* security information on a "need to know" basis. That is to say, unless there is a reason for someone (including fellow investigators) to have the information, it should not be released. The simple fact is that to talk unnecessarily about an investigation that is currently being conducted is unprofessional.

An interesting aspect of the issue of confidentiality is the misconception that applicants (candidates) can be investigated without their knowledge. Management, if requesting a background investigation on a possible executive applicant, should be cautioned that if anything more than a simple record check is done, it may well come to the attention of the person being investigated.

Even the so-called simple record check falls under the federal Fair Credit Reporting Act which guarantees a person's right to privacy. An individual has a right to know of such investigation and the right to challenge any derogatory information about himself. The best strategy is for the company to advise the candidate of the company's policy of conducting background investigations and secure his permission.

Examples of Investigative Effort

Company A, a prominent aerospace manufacturing facility on the East Coast, was burglarized of some $50,000 worth of diodes, a miniaturized electrical component. The burglary occurred on the weekend, and the local police were stumped by the case.

In an effort to establish how the crime was accomplished, the in-house investigators, working with the firm's graphics department people, diagrammed the entire layout of the burglarized area and surroundings. After completing the layout and analyzing the various alarm systems covering the plant, the perimeter fencing, lighting, exterior patrol routes, interior patrols, the reconstruction of each interior patrol through watchclock tapes, and, lastly, taking comprehensive statements from everyone who was in the plant that weekend, it was established conclusively that the burglary had to have been committed by employees.

Although the identity of the culprits was never established, the evidence of internal theft was so irrefutable that the company's insurance carrier paid off the loss under its bonding program. This particular case was the first time the carrier had ever paid on an employee indemnification bond when the actual identity of the employee(s) was unknown.

Company B, a well-known manufacturer and retailer of consumer goods, suffered a $400,000 hijacking of one of its trailers en route from Newark, New Jersey, to Atlanta, Georgia. All the authorities had to work with was the statement of the driver, who claimed he had parked his rig at a diner in Baltimore, Maryland, during his meal break and that when he came out of the diner his rig was gone. The trailer was recovered, empty, the following morning in Brooklyn, New York.

Fortunately, the hijackers were unaware of an odometer on one of the trailer's wheels. The cab odometer had been reset to reflect mileage from Newark to Baltimore to Brooklyn. The untampered wheel odometer, however, reflected substantially less mileage—in fact, it reflected the exact mileage between Newark and Brooklyn. The trailer had never been to Baltimore.

Company investigators determined, through their interrogation of the employee driver, that he had conspired with others in a theft of the trailer's contents. He was driven to Baltimore as part of the cover. The hijacked load was never recovered; on the basis of the employee's confession, however, the company secured a recovery of $390,000 under its fidelity bond.

Company C, a large fashion department store chain located in the Western states, was suddenly victimized by a ring which had compromised the firm's negotiable "scrip" through counterfeiting. Through an informant, the Security Department located a legitimate print shop in a Mexican border town. The printer was being unwittingly used by the culprits, who, posing as executives, initially ordered and purchased seemingly innocent company forms and letterheads. A total of $3,250,000 in ten dollar notes was being processed in the print shop. There was no U.S. law enforcement agency empowered to go into Mexico to investigate the case.

After securing the cooperation of the printer, one company investigator posed as a print shop employee, and surveillance vehicles manned by company investigators were strategically located on the busy Mexican streets for several days. Radio contact was maintained with authorities north of the border. As a result of this effort, four U.S. citizens were subsequently apprehended with full confiscation and recovery of all counterfeit notes.

These cases dramatically illustrate how a company's investigative function can serve the company's proprietary interests and preclude its losses in manners different from those available through normal law enforcement channels.

Summary

Most large private organizations, and many small ones, frequently require investigations different from those available through public law enforcement agencies.

Employee background investigations; internal theft and other employee crime; protection against external attack; the investigation of undesirable behavior, particularly at the management level; corporate liability exposures; the documentation of incidents during labor disputes; and a wide range of miscellaneous situations peculiar to company operations—all of these situations call for an investigative capability which can best be provided by the security organization.

The scope of the investigative function should be clearly defined by management. Outside investigative services often supplement the activities of in-house investigators, although the latter have advantages in terms of lower costs, superior knowledge of company operations and personnel, and the use of resources well-known to management.

Investigative findings should always be confidential. The prerogative of responding to or revealing those findings is that of company management.

Review Questions

1. List six categories of investigations in which the Security Department may become involved.
2. In investigating external criminal attacks, how do the security investigator's objectives differ from those of the police?
3. Give three examples of types of corporate liability cases.
4. Discuss the advantages of using in-house investigators instead of contract investigators.

Chapter 17

Office Administration

In an efficient and well managed organization, general security office administration has six distinctive functions, each of which plays an important role in the success of the total back-office effort. Irrespective of the number of employees available for office duties, the

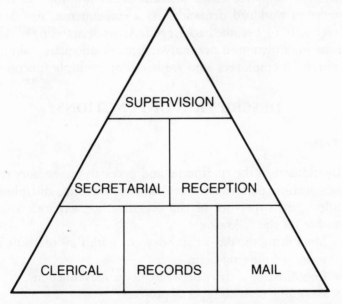

Figure 17-1. The six functions of Security office administration.

211

functions themselves are constant. People may change, but functions do not. Like the placement of stones in a pyramid, each office administration function is unique, critical to the whole, and conspicuous. The functions are:

1) Office supervision
2) Secretarial
3) Reception
4) Clerical
5) Records
6) Mail.

The functions obviously suggest the type of employee required to discharge each function—but that is not to preclude the possibility of joining two or more functions together under one employee, or, conversely, of having two or more employees assigned to the same function. And the interrelationship between functions is most important in the daily office effectiveness and orderly work-flow. Figure 17-2 illustrates possible compacting configurations, depending upon the size of the organization.

The problem with combining functions is that they tend to lose identity; the employee tends to focus or set priorities on personal preference or workload demands. As a consequence, less desirable tasks such as filing become backlogged. An understanding of the risks of combining functions is necessary to provide adequate training and supervision for employees who are handling multiple functions.

DESCRIPTION OF FUNCTIONS

Supervision

In addition to the traditional and necessary supervisory responsibilities such as performance evaluations, training, discipline and scheduling, the supervisor of the overall office activities would be responsible for the following:

- Monitoring the day-in and day-out work flow to insure everything is following plan.
- Coordinating activity between the various functions.
- Assigning tasks and special projects.
- Projecting supply needs, then ordering and controlling supplies.

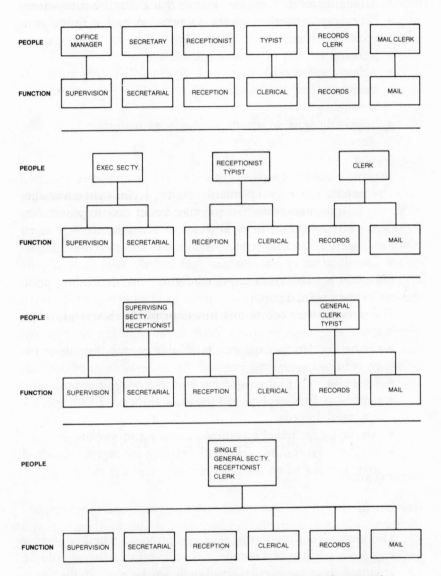

**Figure 17-2. Possible configurations of functions and
employees in office administration.**

- Arranging for maintenance and servicing of office equipment.
- Personally overseeing special projects, including doing such projects himself.
- Insuring security of records and files.
- Arranging for replacement of absent office personnel, re-assigning their functions to others, or filling in temporarily himself.
- Inspecting work to insure standards are maintained.

Secretarial

The secretarial function primarily services departmental management. In a large organization there may be an executive secretary (personally serving the Security Director or Manager) as well as an office manager. Secretarial service usually includes:
- Coordinating appointments.
- Preparing managerial correspondence.
- Assembling and presenting those documents requiring executive signature, such as invoices, expense and travel vouchers, etc.
- Gathering data as requested for budgetary submissions or explanations.
- Screening calls to management and re-routing if appropriate.
- Gathering files, reports and statistics as requested for specific managerial needs.
- Arranging for travel accommodations for management.
- Acting as a communication link between the organization and management when management is not in the office.

Reception

The employee who serves as a receptionist, whether it is a uniformed male officer or a female in regular office attire, should be well-groomed, perhaps even exceptionally so, because of the function's high public visibility and the impression that person makes on the public. If the function is more that of a telephone receptionist, telephone answering training (offered by local phone companies) should be required prior to assignment to the phone board or desk.

People rarely come to or call the Security Department with good news. Meeting a cold, somber or aloof receptionist or hearing an unfriendly voice on the phone only compounds the problems, or it tends to discourage future contacts, even if future contacts would be beneficial to the department. Persons assigned to the receptionist function should be pleasant and courteous in person and have a "smile" in their voices.

Receptionist duties may include:

- Knowing the names and responsibilities of all personnel in an investigative, supervisory or managerial position.
- Knowing the current status of such personnel throughout the day; i.e., at lunch, in a meeting, in court, etc.
- If caller is vague as to the party he wants, determining the nature of the call and routing it accordingly.
- Taking messages for those not in and insuring that messages get to the proper party in a timely manner.
- "Matching up" visitors or callers with the appropriate party.
- Having guests sign visitor ledger and obtain badge, if required.
- Accomplishing sundry light tasks during slow periods, such as matching trailer security seal numbers against manifests, typing index cards, etc.

Clerical

The clerical function is normally filled by a clerk-typist who has a range of office skills up to and possibly including shorthand. This person is the office "generalist," and as the chart in Figure 17-2 suggests, this is a fundamental office function. A sampling of clerical activities includes:

- Transcribing recorded field reports into typed reports.
- "Packaging" files or assembling the contents of a given file into the standard format.
- Logging incoming data on appropriate forms or charts for subsequent compilation.
- Converting compiled statistical data into typed charts or forms for duplication and distribution.
- Processing, sorting and distributing forms and documents as required.

- Preparing index cards on all names processed through Security, with appropriate source reference.
- Typing memoranda, notices, instructions, orders, training materials for duplication and distribution.
- Acting as relief for secretary, receptionist, records clerk and mail clerk.

Records

Records and the records maintenance program may appear to be relegated to a low status, but in actuality the whole record function is the lifeblood of a security organization. To put it another way, the records of a security department really constitute the detailed diary or historical recordings of all security-related events up to the present. So that the reader may better appreciate the scope and importance of this function, the following list represents a sampling of the types of records that may be found in a security records section:

- Company arrest records of all non-employees.
- Company arrest records of employees.
- Employee security terminations.
- Open investigation files on current employees.
- Reports on all burglaries against the company.
- Reports on all arson attacks against the company.
- Reports on all vandalism or malicious mischief.
- Reports on all bomb threats.
- Reports on all obscene phone calls.
- Reports on all major thefts.
- Reports on all suspicious circumstances.
- Reports on all intrusion, waterflow or other fire alarms.
- Reports on all daily patrol activities.
- Reports on all fraud, counterfeit, or impostor incidents.
- Security intelligence files (undercover reports . . . information received).
- Security attendance files.
- Construction, blueprints, etc. on company properties.
- Files on truck schedules and security seal numbers.
- Document files (usually exception or "authorized OK" forms).
- Files of training materials.

- Memo reference files.
- Correspondence files.
- Equipment and resource files.
- Contractor and supplier files.
- Special events files.
- Emergency procedure files.
- Fidelity bonding file.
- Master indices (main index file).
- Accident reports.
- Files on former security personnel.

The foregoing list in no way constitutes a necessarily complete records configuration, nor does it suggest that every security organization requires all such categories. It does illustrate the wide spectrum of possible categories a security department may be required to retain.

Record Filing Systems

To bring order to the records is, of course, the challenge. "Order" means that any given record or file can be located promptly and pulled on demand. Delays or the excuse that "we can't find it" reflect, to some degree, disorder. Order means the material is filed in a systematic way and the system is logical and disciplined.

There is no *one* correct filing system. The design of the system must serve the particular needs of the given organization. The simpler the system, the better. As an example, the entire list above could be filed in a basic alphabetic arrangement, where all burglary files would naturally be filed in the "B" section. An alternative would be to file burglaries under the "C" section for "Crimes" or "Criminal Investigations." For the sake of this presentation, the author will use the straight basic alphabetic system.

The alphabetic system provides the rough or first division of materials. It cannot stand by itself in every case, so a secondary alphabetic or numeric system is necessary. This can best be illustrated by the alpha-alpha system; in filing arrested shoplifters, for example, "S" for "Shoplifters" is the first alpha, and the names of each shoplifter (Abrams, Brown, Cole, Davidson, etc.) form the second alpha.

In the alpha-numeric system, on the other hand, burglaries would be filed under B (the alpha) and an individual case number assigned to each (B-1, B-2, B-3, etc.), the number constituting the numeric control. Depending upon the volume of activity, the system can be "annualized" for increased control and ready statistical information, as illustrated in Figure 17-3.

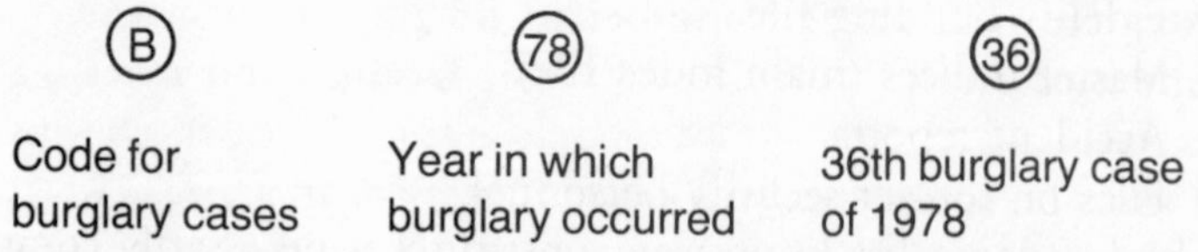

Ⓑ   ⑦⑧   ㉚

Code for   Year in which   36th burglary case
burglary cases   burglary occurred   of 1978

Figure 17-3. Sample of annualized alpha-numeric filing system.

The difference between the alpha-alpha and alpha-numeric method is that the former readily identifies and stores *names of people* and the latter identifies and stores *incidents* in which people are not always known, arrested or identified. If the burglar is caught and known, the master index file (usually a 3x5 index card filed alphabetically) of suspect's name refers to the appropriate file, as shown in Figure 17-4.

DOE, John Howard
BURGLARY SUSPECT
SEE B-78-36

Figure 17-4. Master index file card of suspect's name.

Control Card Ledger

The key to this system is discipline in the controls. A control card ledger must be maintained for all alpha-numeric systems. Usually the card is maintained in front of the controlled section and the next file number is assigned to each case as it occurs. What typically happens is as follows: The investigator is either dispatched to the scene of a reported or suspected burglary, or a security employee comes across the scene of such a crime and calls into the security headquarters for a file number. The records clerk pulls the control card for burglaries and gives the next unused number to the investigator, noting the assignment on the control card by date, location and investigator's name. That case is then an assigned and pending case. The investigation must be complete, "packaged," signed-off (approved by the appropriate supervisor) and filed. Once in file, the control card is checked-off by date and thus accounted for. If the file is then subsequently removed, a note is made on an "Out" card, filed behind the control card, as to who removed the file and on what date. If the file is not in its proper place, yet the control card shows it in and no entry is noted on the "Out" card, someone failed to follow the procedure.

The "Out" card is required for all alpha-alpha and alpha-numeric records, not for simple alpha files, such as correspondence records.

Master Index Cards

An extremely valuable tool in the area of records is the index card in the master index files. Not only is it an alphabetical locator reference card, but it also can serve as a record in and of itself. Incidental information of a derogatory nature can be noted thereon, from a wide variety of sources, and filed for future reference. Local newspapers provide a wealth of information about individuals in the community involved in antisocial behavior or criminal activity. Such information can be extracted, noted on the card and filed away. If at a later time a person by the same name applies for work with the company, the automatic search of the master index (part of the new employee screening process) will disclose the derogatory information. An investigation can determine if the applicant is indeed the same person named on the card. Certainly the fact that the names are identical should never be the grounds for action against the person, but it does give direction to the background investigator.

			OUT		
FILE #	DATE OUT	TAKEN BY		PURPOSE	DATE RETURNED
B-77-41	DEC. 6, 77	JONES		COURT	DEC. 9, 77
B-78-6	MAR. 11, 78	ALTON		SUPPLEMENTAL REPORT	MAR. 12, 78
B-78-6	MAR. 16, 78	ALTON		SUSPECT IN CUSTODY-TO P.D.	

			BURGLARY		
FILE #	DATE ASSIGNED	INVESTIGATOR		LOCATION	DATE IN
B-78-1	JAN. 1, 78	Morlock		Greenridge pump house	JAN. 4
B-78-2	JAN. 3, 78	Sutton		Crenshaw tire center	JAN. 8
B-78-3	JAN. 22, 78	Zeigler		Optometrist office	JAN. 24
B-78-4	FEB. 16, 78	Wagner		Montclair	
B-78-5					
B-78-6					

Figure 17-5. Sample control card and "out" card.

File Control

Security files should never be pulled and released without supervision of the responsible records clerk or supervisor.

Mail

The humble task of processing mail, both incoming and outgoing, receives a lot of attention, and rightfully so. A great deal of business communications is by necessity reduced to the written word and dispatched by mail, whether in-house mail or the U.S. Postal system. Queries, answers, instructions, notifications and a host of other written messages must move expeditiously between given parties. Failure to receive such communications can be costly, cause failures, or at the least be embarrassing. As a consequence, mail failures or undue delays can provoke emotional reactions. Therefore the mail function should include the following:

- One person, whether full-time, part-time or combined responsibility, should be clearly charged with that responsibility. There should be no confusion as to whose assignment it is.
- Periodic daily pick-up from the main company mail distribution room.
- Opening and date stamping of all mail except those envelopes marked confidential.

- Knowledge of who's who, past and present, in the security organization and the company as a whole (to expedite misdirected or improperly addressed correspondence).
- The capability to hand-deliver or hand-post special correspondence.
- Last but not least, an appreciation for the importance of the function. (This can only be achieved by management.)

OFFICE SECURITY

The security office is the repository of a vast amount of information, factual and otherwise (suspicious, not yet verified and still under investigation). All of this data must be categorized as highly confidential and demands security. There are two aspects of security in connection with this information. One is personnel and the other is physical. The people who are hired to work in the office administration area of a security department require precisely the same degree of thoroughness in the background investigation as does any other security employee. The sensitivity of information contained in the files, even in the kind of company not engaged in national defense or other highly sensitive activities, deserves maximum security, even if only from a company liability point of view. The release of information concerning a janitor's discharge due to writing obscene letters to a secretary in the blueprint office, for example, could bring serious and unnecessary repercussions from a civil lawsuit.

Proper physical security requires that all filing cabinets be equipped with a lock as well as a drop bar that runs vertically down in front of the drawers and is secured in place with a padlock during non-office hours.

Additionally, if the office area is in a separate security building, it should be protected by fire and intrusion alarms.

THE OFFICE ENVIRONMENT

The fact that a security department is not a profit-center, or is considered non-productive in terms of the industry or business it serves, in no way justifies hand-me-down equipment or furniture. The security office deserves the same quality of work environment as any other department in the organization, and it is the responsibility of security management to insist on that equality. Office personnel need a clean, fresh, open space in which to work, with good equipment, sufficient light, a place to relax and a general climate which fosters the concept that they too contribute to the overall success of the firm and, as a result, take genuine pride in their department generally and in their work specifically.

Summary

Security office administration includes six key functions. Each is unique and critical. Although it is not always possible that they be performed by different people, it is essential that the specific functions should not lose their identity.

Supervision covers the day-to-day responsibility for the effective functioning of the security office, including planning, provision for supplies and equipment, coordination of work assignments and inspection.

The *secretarial* function provides internal secretarial services for departmental management. The *clerical* role is that of the office "generalist," including typing, filing, processing forms and documents, etc. The *receptionist,* as the department's first point-of-contact with outsiders, should project an image of courtesy and cooperation. The handling of *mail* is also a specific and important responsibility.

Records maintenance is a significant part of the security program. Examples include records of arrest, termination, investigation, crimes of all kinds, reference material, equipment and resource files, etc. Whatever filing system is used, it must provide disciplined control and accessibility of information.

Finally, it is important that principles of security be embodied in the hiring of office personnel and in the internal practice of physical security. The physical environment of the security office should reflect the importance attached to this function by the company.

Review Questions

1. What are the six basic functions of security office administration? Give examples of several of the responsibilities that come under each function.

2. Describe the alpha-alpha and the alpha-numeric systems of filing records.
3. What is the purpose of the control card ledger in an alpha-numeric filing system?
4. What are the two aspects of security for the office administration area of the Security Department?

Part IV
PUBLIC RELATIONS

Chapter 18

Selling Security Within the Organization

Good sense dictates that there is an ongoing need to "sell" the necessity and importance of the security function to the company as a whole. Employees at all levels of the organization must first be made aware of, then understand, and then come to appreciate the fact that the security function is a viable and integral part of the business, whatever that business or industry may be, and as such contributes to its overall success.

Why is there an ongoing need to sell security? Turnover of employees, including those in the managerial ranks, is one reason. A second is a result of the selling effect itself; that is, as security is understood and accepted, its role expands or takes on new internal dimensions (as discussed in Chapter 3) which require new selling.

A final reason is the ever-changing external factors which necessitate change in the security function. For example, race riots appeared on the American scene in the 1960's, followed in rapid succession by civil rights demonstrations, anti-war and general anti-establishment demonstrations, airplane hijackings, executive kidnaping and hostage ransoming, all having a dramatic impact on the private as well as the public sector. The impact in the private sector, of course, fell directly on the security forces. Shifts in security procedures and new security requirements to meet new challenges require selling.

The changes listed above are only some of the most obvious. There have been a myriad of lesser changes in business and industry, causing a security reaction. In retailing, for example, there have been significant changes in criminal attacks. Not long ago retailers were plagued with hide-in burglars. Today that problem has abated, but retailers now face a marked increase in "grab and runs"—thefts in which a culprit enters the store, grabs an armload of hanging merchandise from a stand and runs out of the store to a waiting vehicle with driver, and speeds away. "Grab and runs" were very rare until about 1974. To combat this new theft technique effectively, retail security must sell employees and management on the scope of the problem, what must be done to combat it, and *everyone's* role (not only security's) in that combat strategy.

Selling security, then, is indeed an important security management responsibility.

HOW TO SELL SECURITY

Security First

The security executive cannot sell the necessity and importance of the security function to others if his own people do not understand it. More often than not, the average Security Department employee has a rather limited view of the security function, seeing it only as it relates to his or her particular assignment. They do not see the bigger picture. This "tunnel vision" has a predictable influence on one's attitude, and one's attitude affects one's job performance and relationship with others, in and out of the department.

The single most important aspect of retail security is *shrinkage* or *inventory shortage*. Inventory shrinkage, the difference between the inventory of merchandise on the books and the actual physical presence of goods confirmed by a physical count (inventory count), is the one very visible and tangible measurement of a security department's effectiveness in protecting assets.

In one retail organization, for example, the shrinkage percentage figure, causes of shrinkage, and goals are discussed on posters, in handouts, and in the Security Department's own publication. Yet, at a recent training meeting in the main office and warehouse facility for security officers assigned to that location, not one officer, including those with years of service, could explain the process whereby the company identifies the shrinkage percentage. And not one officer knew what the shrinkage percentage meant in terms of dollars. They were staggered when told that the company, like all major retailers, suffers an annual loss of millions of dollars. When they were told how important they were in the overall efforts to protect merchandise, the light of comprehension seemed to come on. The company's error was in *assuming* the employees understood shrinkage and *assuming* they knew how important their respective jobs were. Today these security officers are thoroughly convinced of the need and importance of the department as well as of their respective jobs.

New Employee Inductions

There is certainly no better opportunity to sell security than that afforded at new employee induction sessions. Not only is there a "captive" audience, but an audience eagerly receptive to information about their new work environment.

Some believe that the presentation on the security organization during the induction program should be made by a member of line management. Even with a prepared script, however, managers tend to deviate from the material, emphasizing those things *they* think are important (which may not be) and omitting information which they feel is better left unsaid because it is distasteful, such as the consequences of internal dishonesty.

Consequently, to insure that new employees are exposed to the information deemed necessary and appropriate, it must be presented either by a security employee or by way of some form of audio-visual media.

The personal presentation is by far the better technique, *if*—and that is an important if—the security employee is a personable, interesting and effective speaker. The higher the rank of the employee making the presentation, the better. Ideally such presentations should be made by the Security Director himself. The further down the chain of command this task is delegated, the lower priority it will be given by the inductees. Thus the very objective of the exposure—to stress the necessity and importance of security within the organization—is defeated.

In a very large organization, spread over a wide geographical area, the Director's personal appearances may necessarily be limited to special events such as the opening of a new facility. Under such circumstances, use of audio-visuals is a good alternative.

Three of the most commonly used audio-visual formats are the slide/tape programs (35mm color photographs projected on a screen with accompanying audio tape); a voice message on audio tape; and video tape. All three can be used effectively to orient, educate and sell security.

Slide/tape programs are inexpensive and relatively easy to put together. This particular medium can be used to explain to employees in a graphic and colorful manner what the Security Department does. Slide/tape programs can tolerably run eight to twelve minutes, long enough to develop an interesting message for general internal consumption as well as for new employee induction sessions.

Audio tape programs should be shorter, probably not exceeding four minutes. For this reason they are more practical for inductee consumption exclusively.

Video tape is unquestionably an exciting medium. One of its advantages is that it tends to personalize the guest or speaker so that people can identify with him or her—a feature not available in slide/tape or audio tape programs. Thus, video comes closest to a live personal appearance.

These media can, of course, be combined in a presentation. Slide/tape or audio programs might conclude with a video-taped interview with the Security Director for added personal impact.

More Audio-Visuals

The use of audio-visuals in selling security is not limited to new employee orientation presentations. One large hotel and restaurant chain uses the media described and, in addition, short motion pictures produced in-house to dramatize security and safety problems and procedures, ranging from the handling of bomb threats to fire prevention.

One retail stores organization has made effective use of an audio tape of an interview between the Security Director and a professional shoplifter, who consented to the interview in return for dismissal of a case pending against him in the local courts. The original tape was a high quality reel-to-reel recording, later reproduced many times on cassette tapes for wide distribution throughout the company.

The shoplifter responded frankly to questions about his trade and skills as they applied to the company. The crook was unquestionably a ham, but his precise answers, his obvious knowledge of the company's merchandising techniques, methods of presenting goods, use of fixtures, floor layouts of individual stores, exact location of stores, one store's laxity in following a given policy compared to another store, what he liked about stealing from this organization and what he feared—all have a hypnotic interest for employees viewing the program.

"Capturing" this thief on tape has made the threat of shoplifting truly credible to the people who can do the most to thwart such activities. He has made literally thousands upon thousands of employees conscious of their role in preventing shoplifting. He has helped to sell the necessity and importance of security.

There are also a wide range of commercially produced 16mm motion pictures and video cassettes aimed at industrial and business consumers. Even films that do not specifically apply to the work scene and security's role there can help sell security—for example, a film on rape prevention presented by the Security Department for the education of female employees.

Executive Orientations

It is as important, if not more so, to deliver the security message to the management team as it is to the line employees. To insure this, one organization requires all new incoming middle-management hires to come through the Security Department for a two-hour orientation (which contrasts with the average one-hour appointment in other departments). Their visit with Security, usually within their first month on the job, is part of an overall company orientation. The new Controller or Unit Manager thus becomes acquainted with department heads and their philosophies. This is certainly not an innovative practice, yet Security is not always included in this type of executive orientation, and it should be.

Consider the impression made on the new executive. He meets the Security Director in the latter's office, where, after light conversation, he is given an organizational overview of security. He is provided with an organizational chart on which he can fill in the names of key supervisors and their phone extensions for his future reference. He is asked about the security function of his previous employer and, using that as a comparison, the Security Director emphasizes the differences, pointing out the merits and virtues of the new company's program over what the new executive is accustomed to. Following that he is introduced to an assistant, who spends time discussing operational practices and problems. Then the executive is introduced to the balance of the department's staff personnel and is given a tour of the security offices.

These new managerial personnel are partially convinced of the importance of security when they arrive, due to the importance attached to the orientation schedule and the two hours devoted to security. There is no question in their minds when they leave the security building that the security function is in the mainstream of the business and has its place in the sun.

Security Tours

Tours of the security facility are a dramatic way to sell security at all levels in the organization. The behind-the-scenes look is intriguing to most people, comparable to the fascination capitalized upon by the television and movie industry in "cops and robbers" entertainment.

To take a class of line supervisors out of their Supervisory Training School and give them a tour of the Security Department usually proves to be a highlight of their entire program. Seeing the proprietary alarm room, the communication center, the armory, the fraud investigators at their desks, the banks of files and indices referred to in background investigations—all this makes a lasting impression on employees.

Bulletins

An important aspect of selling is advertising. The power of a strong ad campaign is well known. Advertising copy has to be directed toward its market, must be interesting, and must have some regularity or consistency in terms of exposure. Given these criteria, the Security Newsletter for Management discussed in Chapter 12 constitutes part of the Security Department's ad program.

This four-page monthly publication not only keeps company management informed of what contributions the security organization makes, it is also used as a source document for meetings and loss prevention discussions.

This type of bulletin is a natural selling and communication tool. People are curious about crime and the unusual (look at your newspapers), and when such events occur in their neighborhood or workplace, their interest is intensified. Unless the dissemination of security events compromises security, why not share interesting aspects with other employees? Doing so highlights the necessity and importance of the security function.

Meetings

Visibility and the opportunity to speak and answer questions about security during company meetings is a powerful way to sell the organization. Because security usually has an impact, to some degree, on every aspect of company life, security has a message for whatever department of the company is holding a meeting.

The objective is to achieve visibility and a piece of the meeting agenda. As a rule the person who is calling or conducting the meeting is receptive to enlivening the agenda, and the change of pace and interest that a Security Department representative brings almost guarantees time. There is always an issue to speak to, depending upon the composition of the group; for example, at a meeting of Personnel Department employees, Security could talk about recent bond and application falsifications, and the importance of Personnel and Security working together to insure that only the highest quality of applicants be brought into the company.

The Security Director and Security Manager should participate in these meetings, but not exclusively. It is important to delegate this function down through all levels of the security organization to the first line supervisor. This not only helps security supervisors grow, it establishes an unofficial "speaker's bureau" and thus greater exposure. If there is reluctance to permit supervisors to speak for the department (usually due to fear they will say something that does not meet with management's approval), then prepare canned presentations and practice giving these talks in your own training session.

Security's involvement in company meetings can take many

directions. In one retail organization, for example, that participation included the following:

- *Regional Store Managers Meeting.* Comprised of store managers and key staff personnel from one region of the company. During this meeting the Security Agent in charge of security within that region asked for and received fifteen minutes on the agenda, during which he reviewed the policy of scheduling fitting room checkers and budgetary considerations connected therewith. Questions and discussion revealed the topic was timely. The Security Agent left the meeting with a sense of accomplishment. Later feedback indicated that the Agent made an impressive presentation and there had been good dialogue.

- *Personnel Managers Meetings.* The Security Director asked for thirty minutes to discuss recent conflicts between Personnel and Security over employee disciplinary decisions. The essence of the message was: "By virtue of our different responsibilities, we are bound to find ourselves from time to time on collision courses. Why collide? Why must we have a Win-Lose relationship? Instead, if the matter cannot be resolved to the satisfaction of both sides, refer the issue up to the next highest level for a decision." It was a positive and constructive meeting, and the Director's time was expanded to sixty minutes.

- *Department Managers Meeting.* The Security Manager met with department managers responsible for high loss areas in the business. He discussed contributing causes and suggested ideas on how they could combat such losses. The meeting was small and there was a great deal of attendee participation. The loss area under discussion was not due to theft but to "paperwork errors," yet Security's presence and interest was a plus; in the department managers' eyes, the Security Manager had stepped outside his traditional role and was assuming a different managerial dimension. He helped sell security because of that, as well as coming across as a pleasant and intelligent person, interested in their problems.

- *Dock Workers Meeting.* A Security line supervisor attended a meeting in the warehouse for dock employees. A video tape of a commercially prepared motion picture on internal theft was exhibited. The security supervisor answered questions following the film. This is a tough situation, to stand up and be willing to take any questions, and the dock workers knew it. Following is an exchange at one of those meetings:

Dock Worker: Is it really true you use spies . . . undercover agents?

Security Agent: That's true.

Dock Worker: And you use them here?

Security Agent: Yes.

Dock Worker: Would you answer this then. How many you got? (a tittering runs through the assembled group)

Security Agent: It fluctuates, but probably right now, somewhere in the neighborhood of thirty.

(almost an audible gasp comes from the group)

Dock Worker: Thirty. Wow! Would you answer this question then. What are their names?

Everyone, including the dock worker and the security agent, howled with laughter. That worker sat down and another stood up and changed the subject with an entirely different question. There was a good feeling about that meeting and it helped, again, to sell security.

Involvement Programs

Programs or activities that bring non-security people into personal contact with Security, with a common goal, tend to cement good relationships.

At one university, for example, students have worked with the Campus Security Department as volunteers. The volunteers are furnished a Security bicycle and two-way radio for shift patrol work. A similar program exists at another campus, where students voluntarily patrol wooded areas of a large Eastern university on their own horses. They provide this service to the security organization in exchange for the facilities for keeping their horses on campus.

Not only does the personal involvement have a positive impact on the individual, but his involvement, if visible to other employees of the company, serves as an example. The logic is simple: if students see other students patrolling areas of the campus, then they realize there *must* be a need for security, and if security is necessary, it is important.

Selling security within the organization sets and maintains a climate of understanding, appreciation, and support for the department's objectives. Some of that support comes in the form of budget dollars.

Summary

There is an ongoing need to make all employees in the company aware of the importance of the security function. Security employees themselves should understand the importance of the security function and of their respective jobs.

Induction sessions for new employees offer an opportunity for presentations by the Security Director or another representative of the department. Audio-visual materials such as slide/tape programs, audio tapes, video tapes and motion pictures can be used effectively.

All newly hired middle-management personnel should undergo an *orientation session* with the Security Department. During this visit the new executive learns about the department's structure and function, meets the staff and tours the security facility.

Bulletins such as a Security Newsletter can be used to inform management of the Security Department's contributions to the company. *Company meetings* offer security representatives a chance to talk about their department's functions and answer questions. Security's relationship with other departments can further be improved by *involvement programs* which bring non-security personnel into contact with Security.

Review Questions

1. Discuss the advantages of having new company managerial personnel come through an orientation session in the Security Department. What are some of the subjects that might be discussed in this session?
2. Discuss how each of the following can contribute to the task of selling security within the organization: new employee induction sessions; company department meetings; bulletins; audio-visual materials; tours of the Security Department.

Chapter 19

Relationship With Law Enforcement

All law enforcement agencies are "security" organizations of one type or another, and many security departments (in the traditional sense) are actively engaged in the purest of law enforcement responsibilities, such as crime prevention, detection, apprehension and prosecution. The key difference between public law enforcement and private security is that law enforcement is a product of and serves the public sector, whereas security is a product of and serves a given segment of the private sector.

Policing, then, is a responsibility of both public and private police. The distinction between the two is found not so much in the organizational responsibility and objectives as it is in the master they serve. There is an absolute necessity for both—and a mutual dependency. The degree of harmony with which the two interface will be affected by many variables, but the relationship can and should be one of effective cooperation rather than friction or competition. In the words of the Report of the Task Force on Private Security, "Ideally, public law enforcement and private security agencies should work closely together, because their respective roles are complementary in the effort to control crime. Indeed, the magnitude of the Nation's crime problem should preclude any form of competition

237

EFFECTIVE SECURITY MANAGEMENT

between the two. Rather, they should be cognizant and supportive of their respective roles in crime control. . . .''*

An interesting relationship existed between the private and public sectors in a mountain community in central California, a town inhabited by employees of a very large utility company headquartered many miles to the south. The company's Security Department was represented in the form of a "resident" Special Agent, affectionately referred to as "Sheriff" by residents for miles around (not only those who were employees of the utility company). The Special Agent was provided with a company car equipped with a radio for communication with the County Sheriff's Department based in the valley below the mountain range. Except for the summer months, there was an unwritten agreement that the "law" was represented by the Resident Special Agent. In matters ranging from accidents to criminal offenses, the Agent responded and subsequently turned the matter over to the county. Although a representative of the private sector, he served the interests of the public sector as well.

Whether the line between security and law enforcement is fine or fuzzy, there is a great deal of movement back and forth over that line. There is a continual flow of retirees leaving the public sector and entering the security industry. There are law enforcement aspirants who launch their careers from the security industry. A growing number of "Security Degree" options are being offered in the Criminal Justice, Police Science and Administration of Justice programs at the community and state university level. All point up the simple fact that the two career paths run parallel.

The strongest link between the two factions is usually at the investigative and administrative levels, and the relationship there is one of mutual respect for the contributions each makes to the successful completion of the task, be it a stolen credit card case or plans for protecting a foreign dignitary. As a rule, the poorest or weakest relationship is at the lowest organizational end of both factions—the police patrolman placing the security officer at the bottom of the "pecking order" (the FBI is best, then Secret Service, State Trooper, City A Police over City B Police, etc.).

Focusing, then, on the relationship at the investigative and

---

*Private Security: Report of the Task Force on Private Security, op. cit., p.19.

administrative levels, let us analyze what the public sector does to assist the private sector, and vice versa.

SERVICES OF THE PUBLIC SECTOR

Provides Information on Individual Criminal Histories

Formerly, this was the most actively pursued aspect of the relationship between the two sectors. The vital flow of intelligence has dwindled to a trickle today, because of right-to-privacy legislation.

The relationship was (and still is, to a lesser extent, today) one of many facets. Depending upon the jurisdiction in terms of state and local laws, and depending on the administration, the records of police departments might be wide open and accessible to security investigators, or "sealed" and officially unavailable. The process of obtaining criminal history information ranged from (1) making a phone call to the Police Records Division and asking for a name check, to (2) actually paying a police officer a pre-set fee for name checks, or (3) securing the necessary check through acquaintances or friends in the police department, or (4) having security investigators in the reserve or auxiliary ranks of the police department so that they had some access to criminal record files.

In the past, whatever was required to gain access to those records was provided, because of the need for the private sector to conduct its own background investigations. The fast food chain which is about to place an employee or prospective employee in a position of trust in their finance division cannot go to the local police and ask them to "clear" the applicant. They would be laughed out of the station. If the police did provide such services (which indeed would be a true service to the business community and certainly would reduce criminal conduct and white collar crimes that affect the community as a whole), they would soon be inundated with requests. So they could not provide such services even before there were legal privacy barriers.

Despite its limited access to criminal histories today, the private sector remains responsible for the consequences of its hiring decisions. For example, a hospital hires an orderly who will work in the children's ward. Local and state police records reveal that the orderly has a criminal history of sexually molesting children. Is it any wonder that security investigators are frustrated when denied information from police files? And in most cases the police are equally frustrated. They know why security investigators seek intelligence and they know how such information will be used, but the growing wall of protection around the offenders has stifled law enforcement's ability to work with the private sector.

The irony of this present condition is as follows: the public sector would not dream of hiring employees into positions of responsibility or trust without examining their backgrounds and assessing any records of criminal conduct—but the private sector is denied that prerogative. Yet the courts have held that the employer is responsible for the conduct of its employees. In a recent case, not yet adjudicated, a security guard raped an employee of the guard company's client. The guard had a criminal history of assaults against women but had never been convicted. There was no way the guard company could obtain that history. (In some states legislation now requires the licensing and fingerprinting of armed guards, but such remedy is limited and still falls short of being the solution.) The civil action brought against the guard company looks as though it will be fruitful. To protect the rights of the man who sought employment as a guard, then, others have paid and will pay a dear price.

The only possible hope this country has today in its efforts against criminal attacks lies in *preventing* crime, not in apprehending offenders after their criminal acts. One technique of prevention is to analyze risks. Without intelligence there is nothing to analyze.

Provides Information on
Possible Criminal Attacks

Through their own system of informers, the police regularly gather intelligence that aids them in their work. It is not uncommon for police to pass on to a security organization information received that specifically affects a given firm or industry. For example, the police department in a large city may learn that the jewelry department of a department store chain with many local stores is targeted for a holdup within the next few days. The greater metropolitan area includes a multiplicity of police jurisdictions, and therefore no single agency could handle the case. Due to the large number of stores, police robbery surveillance teams simply could not be provided.

Acting upon the intelligence gained from the police, the department store Security Department can provide its own surveillance.

On a more general basis, local law enforcement provides security with information on counterfeiting operations, check passing and fraud money order scams, the presence of professional boosters (shoplifters), credit fraud gangs, and a host of other similar intelligence that the security industry needs in its daily efforts to protect the industries it serves.

Provides Traffic Control Support
For Special Events

The type of industry and its influence on vehicular traffic on dedicated streets will determine the relationship with local law enforcement in the area of traffic control. In some cases the only traffic impact may be predictable minor congestions when employees arrive in the morning and leave in the afternoon. This usually can be regulated with signals rather than manpower.

In other cases, such as the grand opening of a new shopping center, the anticipated traffic could be a major concern not only to the merchants, whose interests are defeated if traffic is snarled up and customers cannot get into the center, but to the police as well, who do not want a major traffic jam within their jurisdiction. Thus, the experienced security administrator, anticipating traffic problems at an upcoming opening, will sit down with the local police and outline the probable traffic control needs, usually based on his experience at a previous grand opening in another police jurisdiction. As a rule the police handle the traffic on the dedicated streets and Security provides the manpower for traffic control on the private roadways and on the parking lots.

Poorly coordinated efforts—and they do occur—result in monumental traffic jams (in one case the freeway off-ramp that serviced the street in front of a shopping center was backed up for over a mile and the Highway Patrol had to announce a "Sigalert"). Efficiently coordinated efforts move vehicles expeditiously into and out of the area.

Accepts and Processes Crime Reports

The police accept and process crime reports from firms that have security personnel and from those that do not, as well as from private citizens. Why, then, is this a special relationship with security? If there are no security personnel involved, the police will conduct some inquiry into the facts and details surrounding the alleged crime. Once the proper relationship with security is established, however, the police will accept, *on face value,* the report from a security department. In fact, they will usually accept the completed investigative report from the security department's report form and attach it to their own forms or have the information transcribed from the security report form onto their own form, word for word.

There are two reasons why the police will accept such security reports: (1) They recognize the professionalism (only of deserving organizations, of course) and respect that professionalism. (2) The report's content could be beyond the investigative expertise of the police (e.g., computer manipulation or a cycle variance in accounts receivable).

Coordinates With Security
On Special Enforcement Projects

"Special enforcement" in this context means efforts directed against a general criminal problem, as opposed to investigation which concerns specific suspects.

When the proper relationship is established with the local authorities, a variety of joint projects of mutual interest and benefit may be undertaken. Such projects could include the following:

1. Because of a series of thefts of autos and auto parts, and thefts of packages from cars in a shopping center's parking lot, the security people can set up a surveillance from the roofs of stores and communicate by radio with plainclothes police down in the lot.
2. Because of complaints received that homosexual acts are being solicited in the public rest room at an amusement park or a hospital, a coordinated effort can be effected to detect and arrest such offenders.
3. Because of a series of indecent exposure incidents on the grounds of a local college, the campus security and police can set up stake-out teams as well as set "bait."

Coordinates With Security
On Major or Important Investigations

There are occasions when a criminal case would be impossible to conclude successfully without a coordinated effort of both the private and public sectors. A dramatic example of such a case occurred in Los Angeles. Investigators for a chain of department stores learned that a large number of employees and non-employees were working together in a concerted effort to remove merchandise from the department store's warehouse. Most of the participants were identified, motion pictures were taken of some of the theft activity, and an undercover agent was successfully placed in the midst of the group to provide a flow of intelligence. The department store then went to the local authorities (in this particular case, the District Attorney's office) for assistance.

In a coordinated effort, the following occurred: A small electrical supply and service store was obtained about two miles from the warehouse. It was wired for voice recordings. A panel truck equipped with 16mm motion picture camera was parked behind the store. Two D.A.'s investigators posed as employees in the store and one manned the camera vehicle. Department store investigators secretly marked the kind of merchandise the undercover agent had indicated would be stolen the next day. Through the undercover agent, word was passed to the thieves that there was a new "fence" in the area (the electrical supply store). The department store provided the money to buy the goods. In a short time regular trips were being made to the back door of the "fence," and investigators were buying stolen merchandise marked by other investigators the night before. The transactions were recorded visually by the hidden movie camera as well as audibly.

Results: a grand total of twenty-seven culprits were either indicted and arrested, arrested and referred to juvenile authorities, or, in those cases in which a public offense could not be established, discharged from the company.

A case of this complexity and magnitude could not have been resolved so successfully had it not been for the cooperation between private security and law enforcement. Criminal investigations provide frequent opportunities for this effective interface.

Provides Intelligence
On Radical or Political Activists

Advance information about planned protest marches, demonstrations, rallies—in terms of location, time, who is gathering, their
objectives, the route they will travel, their specific plans (such as tying
up traffic)—is vital information to a security organization which is in
the vicinity of these often socially disruptive activities. The intelligence can be even more crucial to a company which could be the
object of attack—such as a major department store whose travel
bureau sold excursion tickets to the Soviet Union. The Jewish Defense
League demonstrated at the store, disrupting normal operations and
blocking the doorways by sitting down and joining arms. Advance
information allows time for planning a defensive strategy and course
of action.

Good planning based on good intelligence pays off quickly. For
instance, a major department store in Los Angeles was alerted to a
planned Chicano rally at police headquarters, about a half mile from
its downtown store. The store's security force was beefed up. The rally
was dispersed by police, sending hundreds of protestors (and hooli-
gans, who tend to join such activities) fleeing down the main avenues
leading away from the police facility. Groups of these youths broke
hundreds of plate glass windows of stores lining the streets in their
escape route. The department store lost five huge plate glass windows, but because its security personnel were in force, no merchandise was looted from the displays, and plywood panels (pre-fabricated
for just such an event) were quickly erected for temporary security.

The absence of this type of intelligence from law enforcement
agencies would put the security administrator at a marked disadvantage, if not in a hopeless position. And the flow of such information
is directly related to the relationship of mutual trust and respect that
has been established.

Provides Protection
During Labor Disputes

Without question, the Security Department is the enforcement
arm of management. As a consequence of that reputation and profile, its peace-keeping capabilities during a labor dispute are nil.

Only the police can maintain any semblance of order on the picket lines, and even they have their grief because strikers tend to view them as protectors of management and management's property. Security's role is limited to a perimeter defense line on the company property. Police assistance is necessary in terms of keeping the peace, preventing or at least reducing violence against "scabs" or supervisory and managerial personnel, preventing the blockading of access roads, sidewalks and driveways. Without that type of regulatory order, dangerous situations might sometimes escalate out of control.

SERVICES OF THE PRIVATE SECTOR

The security industry's relationship with—or, perhaps more aptly stated, the industry's contribution to—the general law enforcement picture would include the following:

Contributes to the Local
Criminal Statistical Data

The annual Uniform Crime Reports published by the Federal Bureau of Investigation are based on data generated at the local level. The statistical tracking of the number of various types of crimes in a community, the number of arrests, and the number of unsolved crimes serves a number of purposes, including possible budget justifications.

On first observation it might appear that the volume of criminal acts and arrests reflects police activity only. Yet the truth of the matter is that a percentage of these statistics reflect the activity of the security forces in the community—particularly in certain crimes such as larceny. The percentage could be quite small or substantial, depending on the community and the composition of the local police and security forces. The private security force responsible for maintaining order on a large college campus, for example, might often process as many offenders through the "booking" procedure as does the local police agency. In this circumstance, the community served by the private security agency might actually be larger than the one within the jurisdiction of the public law enforcement agency.

The published criminal statistics, then, reflect to some degree the joint effort, but more often than not the independent efforts, of *both* the public and private sector.

It should be pointed out that the Index reflects only "known" crimes; that is to say, crimes known to the police. Actually there is a great deal of crime known to the private sector but for a number of reasons never reported to the police—such as petty shoplifting cases and large internal larceny matters where the best interests of the company are served in a recovery of the loss instead of a prosecution, with its delays, costs, and doubtful outcome.

Provides the Community With
"Tax-Free" Law Enforcement

The Security Department of a single department store in Los Angeles in one year will arrest and prosecute in excess of 2,000 offenders for such crimes as theft, burglary, forgery, credit fraud, counterfeit passing, indecent exposure, and others. Add up all department stores, plus chain stores; add all supermarkets and drug stores; add all the discount stores (including only those retailers who support security forces), and project their combined efforts. Such projection suggests a conservative figure of 50,000 arrests by security forces in the greater Los Angeles area each year, and the number is growing. Bear in mind that these figures apply only to the retail industry (which represents that segment of the private sector most actively engaged in crime/arrest activities). If the burden of that criminal behavior rested on the police, imagine what it would mean in terms of tax dollars!

Provides Liaison Between
Law Enforcement and the Business World

The vast pool of intelligence and resources needed on a daily basis by the police is readily available through the various security organizations serving business and industry. In some cases the intelligence needed is available within the Security Department by virtue of the type of organization—such as an investigator for a telephone company with high technical skill levels needed in some police investigations. In other cases the need can be filled only by a security investigator because the security man has access to intelligence—for example, background information on a present or former employee.

Public sector law enforcement personnel would be the first to agree that many doors would never open without a court order if it were not for the intercession of the security organization. Many companies, by written policy, will not release any information to any governmental agency, but will refer such agency to their Security Department. If the Security Department agrees to the release of information, it is released; if not, it is not released.

Pages could be filled with examples of the relationship between the public and private sector in this area of liaison. The following single example will serve as an illustration, not only of this aspect of cooperative effort, but of the thrust of this entire discussion of the complementary roles of private security and public police.

A young police officer took an elective course in security while pursuing his bachelor's degree in Police Science. A local Security Director taught that class. Some years later, that officer was a homicide detective working on a puzzling death. The only thing found in the deceased's pockets was a cash register receipt. Recalling his earlier studies, the detective called his former teacher. Examination and interpretation of the impressions on that receipt provided a wealth of information that led to solution of the case. The receipt was for the purchase of a specific classification of merchandise on the date of death. It provided information on where the purchase was made in the city (by store identification number), when during that day the purchase was made (by transaction count), and who sold the merchandise to the deceased (by employee ID code). Armed with that intelligence, and following an interview with the employee that was arranged by Security, the detective resolved his case.

The relationship between the private and public sector in this case, as in virtually all cases, served the professional interests of law enforcement specifically, and the welfare of the community generally.

Summary

The functions of private security and public law enforcement often overlap; the two career paths run parallel. The relationship between security and law enforcement is usually one of mutual respect and cooperation at the investigative and administrative levels.

Law enforcement formerly provided the private sector with

criminal history information to assist in background investigations. In recent years, legal decisions on the individual's right to privacy have restricted this service. Police often pass on to security organizations *intelligence concerning possible criminal attacks or civil disturbances.* The police can offer *traffic control support* for special events. The police accept *crime reports* from professional security organizations. The police may assist security in *special enforcement projects* concerning general criminal problems. *Major investigations* often require the coordination of police and security efforts. Police assistance is usually required in maintaining order during *labor disputes.*

The private sector assists law enforcement by contributing *criminal statistical data.* Private security departments arrest and prosecute many offenders each year, providing a form of *"tax-free" law enforcement.* Security departments can serve as *liaison between law enforcement and the business community,* providing intelligence and expertise as needed.

Review Questions

1. Briefly explain six services which law enforcement can provide private security; and three services which private security can provide law enforcement.
2. Discuss the controversy surrounding law enforcement's providing of criminal history information to security investigators.

Chapter 20

Relationship With the Industry

No matter how successful or effective a security function may be or appear to be, there is always room for improvement. There is always a better way. Only the manager isolated from the security community is satisfied with his operation. The need to grow, to reach out for ways to improve, is the mark of a progressive and enlightened security professional.

Where does one reach? The administrator must reach out into the security industry and its vast reservoir of resources and experience. Never before have we better understood the impact of today's shrinking world upon industry. It is a fascinating fact that one administrator's problems in New York City today may be identical to another's in Los Angeles tomorrow. And, perhaps more fascinating, one administrator's solutions to such problems may be better than the other's.

In the private sector, comparable industries or businesses are in a competitive mode, competitors in all things—with the exception of protection. In the case of neighboring universities, one institution competes with the other for academic standing, scholastic achievement within the faculty as well as the student body, athletic prowess, funding, enrollment, percentage of graduates continuing on in graduate studies, etc. Despite that climate of competition, the security heads of those same institutions meet and discuss common problems in a spirit of mutual cooperation, sharing ideas and information for the purpose of improving their efforts to protect their respective institutions.

A department store vigorously competes with its counterpart at the opposite end of the shopping mall in the timing, frequency, merchandise mix and price-points of sales, as well as other gimmicks to attract customers, such as in-store promotions around public figures and drawings for such prizes as new autos or ocean cruises. Competition is so fierce that if one store extends the hours it is open to the public, the other immediately follows suit. Yet, the security people work cooperatively, advising or warning one another of potential shoplifters, bad check passers, or credit card frauds.

The very nature of the security business demands communication and an effective relationship with the industry. The relationship can be divided into four categories: cooperation, participation, contribution, and education.

Cooperation

The cooperative relationship within the industry can be divided into individual cooperation and organized (or structured) cooperation.

At the individual level, to establish and then develop personal contacts with peers is an important, if not vital, dimension of the security professional. Although selecting contacts is usually a highly subjective process, commonality of operations and respect for the individual are important. Once established, a professional kinship grows, allowing an honest exchange and sharing of ideas, opinions, and strategies, much to the mutual benefit of both parties. The key word here is *sharing*. How many times over the years does the typical security manager pick up the telephone and call a counterpart across the country or across town with a request for information or assistance! To be denied that capability would be crippling.

Despite the obvious reasonableness, let alone necessity, of maintaining such relationships with others in the business, some have voluntarily maintained very low profiles and could be called "isolationists." Such a posture is undoubtedly based on the false assumption that one must be a very outgoing, sophisticated and socially charming personality to make professional contacts and friends. Although a good number of professionals are socially comfortable in groups, a surprising number are shy or at least reserved. For many people it is an effort to approach a stranger, even if he too is alone, and strike up a conversation. But people in the security business do it, mostly because they feel compelled to for the very reasons discussed above.

Of course, the closest contacts are usually intra-industry—for example, hospital security people are in closer and more constant touch with others in their own area of specialization than with retail security people. However, there is a great deal of cross-industry cooperation. In a given period of time it is not at all unlikely that a retail security professional would be in touch with the utilities people for information on identity of possible subscribers at a given location; or with amusement park security for information on a former employee; or with airlines security for assistance in moving highly valuable negotiables between cities. If the two security counterparts are known to each other, cooperation (if legal and feasible) can be assured. If they are unknown, the degree of cooperation could be different.

The candid security administrator today who enjoys any degree of success will freely admit that his status is not the result of self-achievement, but rather the result of harmonizing all the input from many sources, including the influence of contemporaries in the industry. In sharing experiences, successes, failures and strategies, we change and grow.

There is a fine line between individual cooperation and organized cooperation, and one tends to weave back and forth over the line. Organized or structured cooperation comes about through associations, societies, or other organized gatherings or relationships of security professionals.

Perhaps an outstanding example of organized cooperation can be found in the retail industry's Stores Protective Association, serving the Southern California area. Comparable "mutual" protective organizations exist in New York City and Washington, D.C.

SPA, as it is commonly referred to, is a non-profit corporation owned by a number of different department and specialty stores throughout the area which sustain the corporation through membership dues and charges levied for services provided. Services include "integrity" shopping, furnishing undercover investigators, furnishing some temporary security personnel, bank check collections, and new hire screening.

The structured cooperation is best seen in the new employee screening operation. All membership stores contribute, on an ongoing basis, the names of all persons arrested for shoplifting and other related retail crimes, and the names of all dishonest employees, to the master index files of SPA. Names of new employees are researched against this immense file, with a predictable percentage of matchups.

This type of formalized industrial cooperation is obviously beneficial. At the time of this writing the northern part of California is exploring the possibility of joining Stores Protective Association, which would not only give stores there a vast resource of intelligence to draw on, but would add to the resources presently available to the companies in the south as well.

Another structured cooperative effort through SPA is their regular bulletin covering successful and attempted criminal attacks against member stores, including physical descriptions, names of suspects if known, methods of operation, vehicles used—all aimed at informing all stores to aid in future prevention or apprehension of the criminals. Additionally, SPA publishes what are called Special Bulletins, warning the various security departments about particular criminals, including their photographs. The bulletins are highly confidential.

The heads of the member stores' security departments meet monthly, usually for a luncheon hosted by one department, and there they discuss the problems of their industry. Their relationship is not restricted to the monthly meeting but is continuous in nature, one calling on the other when the need arises. Personalities always play some role, but the real relationship between these professionals is one of mutual respect and the willingness to share.

The high spirit of cooperation in the greater Southern California area is not uncommon among retail security organizations around the country (or around the world, as far as that goes). Many major cities have a similar, if not organized, retail cooperative relationship. The same is true in other industries.

Participation

The obvious advantage of participation, as evidenced by the Stores Protective Association, goes beyond the narrowly restricted and limited membership of one specialized field within a field (department and specialty clothing stores within the retail industry). Participation in other organizations has other advantages.

Organizations range from local to national, all offering the professional, at one time or another, something in the way of growth. Every security supervisor, manager, or director, irrespective of his particular industry, has a variety of professional organizations he may join as a member, visit, or infrequently partake of—all aimed at enhancing his knowledge and his personal or departmental efficiency. A retail security executive in Los Angeles, for example, would have the following available to him:

1. Stores Protective Association
2. Special Agents Association of Southern California
3. Chief Special Agents Association of Southern California
4. National Retail Merchant's Association
5. American Society for Industrial Security.

This list excludes a variety of law enforcement associations (such as the State Peace Officers Association), training or educational organization membership, other national groups of limited interests (such as polygraph associations), the Atkins group (a number of department stores with a common interest), and corporate gatherings.

Participation in the groups at the top of the listing provides the greatest day-in and day-out impact on individual and departmental efficiency. As one moves down the list (inserting the excluded organizations where appropriate), the impact becomes proportionately less —but important nonetheless.

An overview of participation in those listed organizations follows:

1. *Stores Protective Association.* Discussed above.
2. *Special Agents Association of Southern California.* This is a local association of security people of all ranks, with an active associate membership of service and security supply people. The emphasis is on retail in general; i.e., not only department and specialty stores, but food, drug and discount as well. In addition to the fraternal interests, emphasis is placed on outside speakers discussing topics of mutual interest, such as opinions of the bench, pending legislation, governmental practices affecting the industry, and management practices. Concerns are local in nature, not national.

3. *Chief Special Agents Association of Southern California.* This group is restricted to security chiefs of major companies, with limited associate membership of number-two people. The association is fraternal, with emphasis on strong liaison with top local law enforcement officers. A significant and very noteworthy contribution to the security community is the publication every other year of a directory of all security and law enforcement organizations in the area and their key personnel. The directory is the only one of its kind and is coveted by all law enforcement as well as security personnel.

4. *National Retail Merchants Association.* This is a security group comprised of department, chain and large specialty store top security administrators who meet annually to exchange retail security problems and solutions, on the broadest possible basis. What is happening in New Orleans could happen in Seattle. Stores in Philadelphia, working with the Chamber of Commerce, are involved in a community-wide effort to reduce shoplifting. That program could work in San Francisco.

5. *American Society for Industrial Security.* The epitome of the profession, this group serves the entire security industry in terms of membership and interest. Local chapters bring together professionals from all industries so that there is a cross-fertilization of ideas and experience. Emphasis is on upgrading the industry by establishing standards and educating members through professionally presented training seminars, workshops and annual national meetings.

The organizations named represent a mix of specific, local and general industrial interest. The reader, student or security practitioner can and should substitute and add other relatively comparable organizations, in keeping with his particular interest or field.

Clearly, participation in security-related organizations and their programs brings to the security professional an ever-increasing array of knowledge, insight and strategy that is otherwise unavailable. Failure to pursue involvement and active participation only brings stagnation and a truly narrow approach to the business of protecting one's company.

Contributions

One facet of participation is contribution. It is a rare man or woman indeed who can participate (belong to and attend) without making some form of contribution. To sit at a round-table discussion and comment does constitute contribution. One-sided participation would be selfish, denying others what the non-contributing attendee himself seeks. Participation, if open and natural in exchanges, serves others in some small measure.

But that amount of contribution, important as it is, is not sufficient for the welfare of the security industry. Where would the state of the art be today without textbooks written by security professionals and articles submitted to trade magazines? Someone must give direction to our educational institutions about appropriate curriculum for students now pursuing degrees in security management. And who is talking to the community, to service organizations, and to us about the business deemed so important in the economic health of private industry and business?

Our relationship to the security industry in general, if it is to be constructive and positive, if it is to help upgrade, if it is to continue to strive for professional recognition of our chosen careers, must have a contributory dimension. Untapped talent surrounds us in our business. A brief examination of some significant areas of contribution, such as book authorship, article authorship, public presentations and involvement in the educational process, is in order.

Book authorship—Healy, Curtis, Hemphill, Green and Farber, to name a few, have taken the time and energy to make a contribution to our industry. As small as their own perceptions of contribution may have seemed, they have been significant and substantial. In most instances, those in the industry who have made a literary contribution are not primarily authors. Rather, they are security professionals, willing to share their point of view, their opinions, and their experiences. In view of the past, present and projected growth of the industry and the increasing number of colleges and universities offering this discipline, security needs more people who are willing to contribute their experiences by reducing them to paper for subsequent book publication.

Magazine article authorship—Essentially the same thing can be said about article authorship as book authorship. At a recent meeting of retail security executives, one participant was overheard making relatively uncomplimentary remarks about the quality of articles in one of our important trade magazines. How much easier it is to criticize than to contribute! Trade magazines are but a vehicle through which the trade can speak. The publisher, editor and magazine staff are publishing professionals, not security professionals. The quality of trade journal content is a reflection of the quality of the trade. The security administrator who considers himself talented but fails to share that talent obviously fails to contribute to the best interests of his chosen profession . . . and/or he fears the bright light of public scrutiny which always brings criticism.

Public presentations—public presentations, which include in-house talks, presentations before the industry, talks before other trades, or luncheon speaking, tend to give the security function high and positive visibility to the non-security audience and add to the professional growth of those in a security audience. Appearing before the local Rotary Club has little to do with the speaker's relationship to the security industry, except that through his presentation he may generate good will for the security cause, with some rippling effect. The same is true with in-house talks and presentations made to other disciplines, such as a Controller's Association.

The real contribution comes in making a presentation to the security industry. No matter how experienced or learned a man may be, there is something new to learn. We learn from others, through their willingness to share—if not in book or magazine article authorship, then at least in their personal presentations. For a man who is considered a leader in the industry, or an administrator with a successful track record, to refuse to make a presentation before the industry is unfortunately selfish. Security professionals have an obligation to contribute to the cause, for the good of all. They cannot in good conscience sit back and rest on their laurels.

Involvement in the educational process—Although not obligatory, teaching in the Security Administration field is a very marked contribution to the industry. Not everyone has teaching skills, of course, so opportunities in this area are limited. However, there is an increasing call to service in terms of participating in college curriculum advisory committees now forming in the path of the burgeoning security programs at the college and university level.

Education

The last important category of relationships with the industry is primarily self-serving, but in a positive way. The sum total of an active relationship of cooperation, participation and contribution is self-education. Keeping abreast of the industry, in terms of new technology, new case law, new legislation, innovative concepts, new trends, is as important to the security practitioner and administrator, if not in some cases more so, as are the daily operations of the organization.

For those security professionals who failed to complete their college education, now is an ideal time to return to school and earn their degree. Security Administration programs are being established all across the country in response to a demand for formalized education in the security field.

In this business one cannot afford to sit in the backwaters and permit the rest of the industry, in all its vastness, to stream by. Time was when change came slowly in the security profession. Today we are truly in a state of "Future Shock." Changes are rapid and accelerating. Technical advances in electronics, communications, and alarm capabilities alone are almost overwhelming.

Reading is but one source of information. Discussion, explanation and demonstration are needed to supplement the written word.

The demands of the business require good communications and a participatory relationship within the industry. And this is a two-way street.

Summary

In a continuing effort to improve his operation, the security administrator must look to his peers in the security industry.

Cooperation within the security industry takes the form of individual and organized efforts. Personal contacts among peers allow an exchange of ideas and strategies. Organized cooperation occurs through the activities of formal associations of security professionals.

Security professionals have the opportunity for *participation* in organizations ranging in scope from local to national. Participation usually includes some form of *contribution*. Contributions of the security professional to his field can include book and magazine article authorship, public speaking, and teaching.

Cooperation, participation and contribution form an important part of the *education* of the security professional in his effort to keep abreast of constant changes in his field.

Review Questions

1. What are the advantages to the security manager of active involvement in the security community?
2. Find out what professional organizations are available for security managers in your area to participate in, and what their activities are.

Chapter 21

Community Relations

A security department's involvement or participation in community relations activities is directly related to (1) the public's need for information, (2) the need to inform the public, and (3) security management's receptivity to such public exposure.

One might argue that there is really no difference between "the public's need for information" and "the need to inform the public." The distinction is this: the public's need is a demand and a need *they* identified, whereas "the need to inform the public" is first identified by the company.

An example of the public's need for information might well be a community's concern about the safety of a newly constructed nuclear power generating station. The utility company's Security Director could be an effective representative and speaker at local civic and business organization meetings.

In some cases local concern has taken the form of organized demonstrations and outright hostility. When business or industry moves into a new community, they will meet with some resistance. Today there is much concern about environmental impact, which includes everything from atmospheric emissions, traffic congestion, noise pollution, architectural blight on the landscape, down to the killing of old oak trees. Reaction to such issues is often emotional. Local government is usually sensitive to the feelings of its constituency, and actions by local government such as finding fault with construction inspections, making unrealistic demands to comply with local codes, and delaying the issuance of permits and approvals can often be traced back to an ill-informed public.

Certainly senior management of the company must be visible and carry the brunt of a program aimed at establishing good community relations. The question is, does security have a role here? Indeed it does, primarily in the form of a security executive who can first establish a relationship with key local officials in public safety and law enforcement departments and then, through those officials, with community groups. Such a spokesman, if he is involved in the new business project from its early stages, can help allay fears and at the same time build good will for the company—depending, of course, upon his public speaking presence and skills.

The need to inform the public is a challenge to the retail industry, for example, with its constant problems of shoplifting, particularly juvenile shoplifting. For security representatives to go out into the community and talk about the personal consequences of theft and the impact of losses on retail prices plays an important role in educating an otherwise ignorant and apathetic public.

Security management's response (as a whole) to these needs is usually traceable to the head of the security organization. For example, if the Director of Security is gregarious, outspoken and a confident public speaker, then the department will usually enjoy a high profile in terms of community relations. Not only will the Director have public exposure, but others on the staff will be encouraged to represent the organization.

On the other hand, if the Director is uncomfortable or dislikes public exposure, he will avoid it altogether or at least minimize it, regardless of the public's or the company's needs. And it usually holds true that if the Director maintains a low profile, so does his staff. It is simple for the Security Director to avoid or minimize community relations activities; he just passes the need or the challenge to a peer in the industry, to some other Director, or to another executive in the company—someone on the legal or operations staff, perhaps, or the public relations people. It is a rare but strong Director of Security who will refer such public exposure to a subordinate, because the subordinate's high profile and recognition for his activities tends to be threatening.

Ideally, the Director will have public poise, will respond positively to the needs of the community and the company, will be supportive of his staff's participation in community events, and—by virtue of his sensitivity to his own department's role—*will actually go out and help identify needs that can be addressed through a good community relations program.*

Community relations, for the sake of this work, refers to

- Public speaking engagements
- News press appearances or interviews
- Radio interviews
- Appearances on television
- Participation in community-oriented projects
- General public contact.

Public Speaking

The luncheon speaker, after-dinner speaker, guest speaker between the ladies' business meeting and "social hour," the classroom or auditorium speaker—each is very much part of the American scene, not only in his very presence as a vital part of the social context in which he appears, but also in the trappings and traditions that surround the speaking engagement. The process begins with the advance announcement of the program and build-up of the speaker's credentials and his topic. It continues as seating arrangements at the head table are made, as the speaker sneaks last-minute looks at his cue cards or typed text, as he is introduced by the program chairman and is greeted by applause. His opening comments are solicitous of a humorous response to "break the ice." Later comes the final, second round of applause following the presentation, expression of appreciation, and the inevitable little circle of people who surround the speaker after the meeting has been adjourned. All of this is somehow not only a natural, but an *expected* and secure process in our culture.

If the speaker does poorly—and many do—it is so noted and quickly forgotten. What was most important was that the *process itself* was right. Conversely, if the speaker did well, there tends to be a lingering sense of good will about him, his message and his organization. In short, there is little risk in public speaking in terms of ill will, but there is a great opportunity for genuine good will.

As stated earlier, the need of the public (or any segment thereof) for information, and the company's need for the public to know, represent the two *legitimate* reasons for a public speaking appearance on the part of a security executive. Something less than legitimate, then, would be an appearance where there is no need to know. That could well be a church's Ladies' Auxiliary or the local PTA, depending upon the prospective speaker. For example, what is the legitimate need for a hotel security presentation to a club of Vietnam Veterans confined to wheelchairs? Perhaps one could make a case for the presentation on the basis that such vets do travel from time to time, and security tips could be helpful.

More importantly, if an organization calls on, invites or asks for a presentation, that request in and of itself represents a need that can be filled. More often than not, filling that kind of need automatically brings about a great deal of good will.

Good will, then, is the common denominator and underlying theme, not only in public speaking appearances, but in all community relations activities, although the specific objective of any one such activity is informative in nature. Rare is the company that is not interested in good will. The Security Department, by virtue of its unique role and image, is in a position to generate good will for the organization as a whole, if it is willing to.

Some administrators called on to speak worry unduly over the topic selected by the requesting group because they feel inadequate to speak on that specific topic. Guest speakers should understand that the topics requested are usually only suggestions. The group frequently is uncertain what they want. The negotiability of topics gives the prospective speaker latitude to direct the topic and his remarks to areas in which he is more comfortable or which are even more appropriate to the occasion.

Ideally the presentation should be tailored to appeal to the audience and have some constructive application for that audience—even if the presentation is one of good will only. In the case of the veterans in wheelchairs, the hotel security administrator could not only entertain but could provide the audience security and non-security related advice concerning hotel accommodations and services for the handicapped.

This is an important point: the security executive must recognize that he is a spokesman for his industry, be it hotel management, transportation, manufacturing, etc., as well as a security spokesman. To talk in terms of the industry, as a member of its management, adds polish and dimension to his presentation befitting his professional position. Gone are the days of the cigar-chewing security chief who did not know—or, for that matter, care—what the company's objectives were, what they were doing or where they were going.

Table 21-1 suggests possible speech topics for a number of different security functions that might be appropriate for four different types of audiences. Needless to say, the list of security functions, the variety of audiences, and the wider variety of topics could consume pages.

News Press Appearances or Interviews

The author's experience with the press has been that if you are wrong, the press will insure that the public is clearly so informed, and if you are right they will serve as a powerful ally. The distortions which do creep into newsprint tend to be impartial—that is, there is as much favorable distortion as there is unfavorable.

Response to requests from the press for specific news, general news or feature material is, or should be to some measure, controlled by the public and press relations department of the company, which can aptly serve as a buffer between Security and the press if need be. A sound practice and policy is to require that all press contacts be cleared through the press relations people first. This accomplishes two things: (1) the people charged with the release of information to the press officially sanction the subsequent dialogue between Security and the press and the release of information, and (2) such sanctioning gives Security natural "protection" against criticism that can be inspired by distortions or by releases that come as a surprise to management.

Press releases can include the appointment of a new director or manager, significant criminal attacks, significant successes, or unusual achievements (such as winning in a national or local security competition).

Feature articles on a security department and its responsibilities provide interesting reading and favorable publicity for a company, particularly if slanted as an "inside look" at protection.

	Local Business and Civic Groups (e.g., Rotarians)	School Civics Classes and PTA	Church Groups	Professional Groups (e.g., Controllers Association)
Hotel Security	• Orientation of the industry • Interesting loss statistics • Humorous anecdotes • Tips for the businessman who travels	• Career opportunities in hotel management • Humorous anecdotes • Examples of a hotel detective's day's work	• Combination of the civic and school presentation	• Case studies of internal losses • Systems modifications resulting from internal loss cases
School/Campus Security	• Need for community support for school bonds • Security costs • Examples of loss	• Vandalism problems and how they affect the quality of school life • Why the need for security	• Overview of security problems and the need for moral leadership from the church	• What's new in combating vandalism and losses, in terms of strategy and hardware
Aerospace Security	• Discussion of possible application of security technology to local businessman's problems	• A look into the future • Interesting cases and problems • Career opportunities in the industry	• As the church's goal is one of peace . . . so is the industry's • Examples of attacks against that effort	• Unique problems or demands connected with present or new contracts and projects
Utilities Security	• Security measures aimed at protecting resources, the community, and the consumer	• Youth's role in conserving resources • Examples of youth's acts of vandalism and attacks	• Importance of home and church's influence in conserving resources and energy • Example of problems	• Security auditing procedures • Examples of defalcation case in payment processing center
Retail Security	• Discussion of threat and problem of internal dishonesty and ways for small businessman to detect and prevent same	• Statistics on juvenile shoplifting • Consequences of theft	• Overview of shoplifting and how the losses affect prices • Need for leadership in setting moral standards and values	• Examples of recent credit and EDP frauds and prevention efforts in those areas
Transportation Security	• Industry losses and impact on insurance and freight charges	• Transportation's role in community and economy • Attacks against the industry • Consequences of attacks	• General information, similar to that given to the schools	• Hijacking frauds and techniques to detect such frauds

Figure 21-1. Suggested speech topics for various audiences.

Radio Interviews

So-called "talk shows" or "spot" commentaries on radio can serve in essentially the same way as public speaking engagements—i.e., informing the listening public of those things they need to know about security. One successful technique has been the recording of a retail security professional's remarks while talking with a radio reporter. One-minute sections are then extracted from the pre-recorded interview for hourly airing in support of a community's anti-shoplifting campaign during the holiday season.

Radio as well as television exposure depends a great deal on the size of the community and the sophistication of network and local broadcasting programming, coupled with the community's need for information or the considered human interest aspect of a given security department.

The important thing here is for security managers to recognize the potential of these two important community communication media, and to seize on those opportunities when they present themselves.

Participation in Community-Oriented Projects

Some community projects, whether sponsored by the Junior Chamber of Commerce, local government or the efforts of a private organization, deserve if not demand support and participation of security professionals. An example would be a community-wide campaign to reduce auto thefts by encouraging drivers to lock their cars.

Depending upon the campaign's organizing group and contact points in the community, the security executive could be the intermediary between the campaign and senior management of the company. Consequently he would be instrumental in obtaining funds for the effort. Ideally security would be in this position.

Further, those security departments which have public parking facilities, such as amusement parks, hospitals, museums, sports events facilities, shopping centers, etc., could help the cause by distributing literature or notices supporting the campaign on vehicles parked in their areas. Such notices could be in the form of 4-inch by 4-inch red cardboard octagonal "stop signs" that say "Stop Auto Thefts . . . Please Lock Your Car," or some similar message.

Matching bracelets.

Shoplifters get something for nothing.

Figure 21-2. Example of a community anti-shoplifting campaign.

Such programs also provide the opportunity for radio and TV interviews with security professionals supporting that campaign. Such interviews provide excellent visibility for the company and demonstrate its concern for and support of community-wide programs that really benefit the public.

A recent community-wide project in the Los Angeles area was the preparation and distribution of a booklet, entitled "Probable Cause," on a new state anti-shoplifting law. Funding for the publication of this booklet came from four major department stores, although the real benefactor of the publication were the small or independent stores, which have little or no security. Acknowledgment of such support was printed on the back of the booklet.

Other community as well as state-wide anti-shoplifting campaigns have marshaled the support and expertise of retail security executives, with, in some instances, a profound impact on the general public—certainly a positive contribution to the community as a whole.

Again, funding, expertise, distribution, personal appearances, radio and/or television interviews are all viable ways to participate and contribute to community projects that not only serve the best interests of the community, but of the Security Department and company as well.

Figure 21-2 is a further example of a community project.

General Public Contact

Perhaps the most overlooked area of good community relations is the day-in and day-out Security Department contacts with the general public. Courtesy, good grooming, business demeanor, and the individual attitude of each member of the department can—and do—have an impact on the company's image and reputation of community good will.

Summary

The Security Department's involvement in community relations activities is related to (1) the public's need for information, (2) the company's need to inform the public, and (3) security management's receptivity to public exposure. Ideally, the Security Director will respond positively to the needs of the community and will take an active part in identifying needs that can be addressed through a good community relations program.

Methods by which the Security Director can inform the public and create good will include *public speaking appearances; press appearances* and interviews; *radio interviews;* and *participation in community-oriented projects* such as campaigns to reduce auto thefts and shoplifting.

The Security Department's day-to-day contact with the public also contributes to the company's image and good will.

Review Questions

1. List six types of activities which enable the Security Department to establish a good community relations program.
2. Suggest several possible topics for a security manager addressing a local civic group; a high school class; a women's club.
3. Give several examples of community-oriented projects in which security might participate.

APPENDIXES

Appendix A

EMPLOYEE PERFORMANCE EVALUATION

HOURLY AND WEEKLY RATED
NON-SUPERVISORY PERSONNEL

NAME _____   DATE OF RATING _____

JOB TITLE_____   ON PRESENT JOB SINCE _____

EMPLOYEE NUMBER _____   SERVICE DATE _____

INSTRUCTIONS: Read the definitions under each factor listed below and check (✔) the box that best describes this employee's overall performance for the past year. To the right of the definitions are various elements for each factor, again check as appropriate. Any BAS rating must be explained in the comments sections.

RATING CODE DEFINITIONS

(O)	OUTSTANDING:	Performance of extraordinary or rare nature. Consistently exceeds normal job requirements. Makes substantial contributions to the success of the department.
(AAS)	ABOVE ACCEPTABLE STANDARDS:	Performance that frequently exceeds normal job requirements. Makes definite contributions to the success of the department.
(MAS)	MEETS ACCEPTABLE STANDARDS:	Performance that meets normal job requirements. There is no evidence of any major deficiency.
(BAS)	BELOW ACCEPTABLE STANDARDS:	Performance that is frequently below normal job requirements. Evidence of major deficiencies. Improvement is required to meet job requirements.

271

PART I

1. KNOWLEDGE OF JOB

The understanding of basic
fundamentals, methods and
procedures of the job.

A. KNOWS PROCEDURES
B. LEARNS WORK QUICKLY
C. KNOWS EQUIPMENT AND FORMS
D. KNOWS WHY THINGS ARE DONE

O	A	M	B

O	AAS	MAS	BAS

COMMENTS:_____

_____

2. QUALITY OF WORK

Grade of acceptable work
compared to what might
reasonably be expected.

A. ACCURACY OF WORK
B. THOROUGHNESS OF WORK
C. NEATNESS OF WORK

O	A	M	B

O	AAS	MAS	BAS

COMMENTS:_____

_____

3. QUANTITY OF WORK

Volume of acceptable work
compared to what might
reasonably be expected.

A. OVERALL VOLUME OF WORK
B. CONSISTENCY OF OUTPUT
C. EFFORTS TO IMPROVE OUTPUT
D. UTILIZATION OF TIME

O	A	M	B

O	AAS	MAS	BAS

COMMENTS:_____

_____

4. ADAPTABILITY

Quickness to learn new
duties and to adjust to new
situations encountered on
the job.

A. ADJUSTS TO NEW SITUATIONS
B. QUICK TO LEARN NEW DUTIES
C. FOLLOWS COMPANY POLICY

O	A	M	B

O	AAS	MAS	BAS

COMMENTS:_____

_____

5. ATTITUDE   O A M B

The interest, enthusiasm
and cooperation shown in
the work, in the company,
and with associates.

A. INTEREST IN WORK
B. COOPERATION
C. RESPONDS TO CONSTRUCTIVE
   CRITICISM
D. RESPONDS TO TRAINING
E. ALWAYS DOES HIS BEST
F. HELPS FELLOW EMPLOYEES
G. KEEPS SUPERVISORS WELL
   INFORMED
H. RESPONDS TO CHANGES

O   AAS   MAS   BAS

COMMENTS:_____

6. DEPENDABILITY   O A M B

The ability to work without
close supervision. Accuracy
and follow through on
assignments without
constant checking.

A. FOLLOWS THROUGH ON
   ASSIGNMENTS
B. EFFECTIVE UNDER PRESSURE
C. FOLLOWS INSTRUCTIONS
D. ACCURACY IN HANDLING
   PROCEDURES WITHOUT
   CONSTANT CHECKING

O   AAS   MAS   BAS

COMMENTS:_____

7. JUDGMENT   O A M B

Ability to decide course of
action when some choice
can be made.

A. PROPER ATTENTION TO DETAILS
B. REASONING IS SOUND AND
   CONSISTENT
C. TAKES PROPER AMOUNT OF
   TIME TO CONSIDER FACTS AND
   THEIR APPLICATION

O   AAS   MAS   BAS

COMMENTS:_____

8. INITIATIVE   O A M B

The ability to perform
assigned jobs in a self-
confident, eager manner
without detailed instruc-
tions.

A. EAGER TO IMPROVE OWN
   PERFORMANCE
B. TACKLES DIFFICULT JOBS
C. SEES THINGS TO BE DONE
D. INQUISITIVE

O   AAS   MAS   BAS

COMMENTS: _____

_____

9. CUSTOMER SERVICE (As Applicable)

	O	A	M	B

Alertness to, acknowledgment of and interest in the customer.

A. IMMEDIATE APPROACH AND ACKNOWLEDGMENT OF CUSTOMER
B. RECOGNITION OF WAITING CUSTOMER, IF BUSY
C. GRACIOUS, COURTEOUS AND ATTENTIVE
D. GIVES ALERT AND INTELLIGENT SERVICE

O	AAS	MAS	BAS

COMMENTS: _____

_____

10. MERCHANDISE KNOWLEDGE (As Applicable)

	O	A	M	B

Well informed on entire stock in department.

A. VOLUNTEERS MERCHANDISE INFORMATION
B. GIVES ACCURATE INFORMATION
C. USES FASHION AND ADVERTISING INFORMATION
D. DEMONSTRATES USE AND CARE OF MERCHANDISE
E. INQUISITIVE

O	AAS	MAS	BAS

COMMENTS: _____

_____

11. SALES PERFORMANCE (As Applicable)

	O	A	M	B

Overall selling performance.

A. VOLUME
B. ATTITUDE
C. SUGGESTIVE SELLING
D. CLOSING THE SALE

O	AAS	MAS	BAS

COMMENTS: _____

_____

12. APPEARANCE

	O	A	M	B

The overall impression given to the customer. Neat and businesslike or sometimes careless and untidy.

A. APPROPRIATE FOR THE JOB
B. IN GOOD TASTE
C. WELL GROOMED

O	AAS	MAS	BAS

COMMENTS: _____

13. ATTENDANCE
A. NUMBER OF DAYS ABSENT_____SINCE_____
B. NUMBER OF OCCASIONS ABSENT_____SINCE_____
C. OVERALL DEPENDABILITY AS RELATES TO ATTENDANCE
   AND TARDINESS

O A M B

COMMENTS:_____

PART II   NAME_____

SUMMARY RATING

Overall job performance and contribution to   COMMENTS:
the success of the department.

OUTSTANDING

ABOVE ACCEPTABLE STANDARDS

MEETS ACCEPTABLE STANDARDS

BELOW ACCEPTABLE STANDARDS

PART III
1. DISPOSITION

_____IMMEDIATELY PROMOTIONAL TO:_____

_____PROMOTIONAL IN _____ MONTHS TO:_____

_____RECOMMEND TRANSFER TO: _____

_____LEAVE ON PRESENT ASSIGNMENT _____

_____PLACE ON JOB PERFORMANCE CAUTION

          1   2   FINAL

OR WARNING   _____TOO NEW TO SPECIFY

2. RECOMMENDATIONS FOR TRAINING OR OTHER ACTION: _____

_____

_____

3. REMARKS: _____

_____

_____

_____

RATER	DATE	SENIOR RATER	DATE	PERSONNEL MANAGER	DATE

PART IV

NOTE: Do not conduct interview until Senior Rater and Personnel Manager have signed this form.

1. REPORT OF INTERVIEW _____

_____

_____

_____

_____

_____

_____   _____
DATE OF INTERVIEW   SIGNATURE OF INTERVIEWER

Appendix B

SECURITY VULNERABILITY SURVEY

Facility _____ Survey Date _____

Address _____ Facility Manager _____
Telephone No. _____

I. GENERAL FUNCTION

Leased
Owned
_____

No. Employees Assnd. _____

Operating   Weekdays   Saturday   Sunday
Hours:

                Opens _____   Opens _____   Opens _____

                Closes _____ Closes _____ Closes _____

Address & Phone of Police Jurisdiction: _____

Area Evaluation: _____

II. BUILDING & PERIMETER

_____ 1. Type of construction?

_____ 2. Door construction (hinges, hinge pins, solid core, etc.)?

_____ 3. Total number of perimeter entrances?

_____ 4. Are all exits & entrances supervised?
        If not, how controlled?

_____ 5. Are there perimeter fences?
        Type?
        Height?
        Distance from bldg.?
        Cleared areas?
        Barbed wire top?
        Roof or wall areas close to fence?

_____ 6. Are there any overpasses or subterranean passageways?

_____ 7. Height of windows from ground?
        Adequately protected?

_____ 8. Any roof openings or entries?

_____ 9. Any floor grates, ventilation openings?

_____10. Any materials stored outside bldg.?
        How controlled?

_____11. Adjacent occupancy?

        Comments:

III. VEHICULAR MOVEMENT

_____ 1. Is employee parking within perimeter fence?

_____ 2. Are cars parked abutting interior fences?

_____ 3. Are cars parked adjacent to loading docks, bldg. entrances, etc.?

_____ 4. Do employees have access to cars during work hours?

_____ 5. Vehicle passes or decals?

_____ 6. Are guards involved in traffic control?

        Comments.

IV. LIGHTING

_____ 1. Is perimeter lighting provided?
        Adequate?

_____ 2. Is there an emergency lighting system?

_____ 3. Are all doorways sufficiently lighted?

_____ 4. Is lighting in use during all night hours?

_____ 5. Is lighting directed toward perimeter?

_____ 6. Is lighting adequate for parking area?

_____ 7. How is lighting checked?

_____ 8. Is interior night lighting adequate for surveillance by night guards (or by municipal law enforcement agents)?

_____ 9. Are guard posts properly illuminated?

Comments:

V. LOCKING CONTROLS

_____ 1. Does the facility have adequate control and records for all keys?

_____ 2. Is a master key system in use?

_____ 3. How many master keys are issued?

_____ 4. Are all extra keys secured in a locked container?

_____ 5. Total number of safes?

_____ 6. Last time combination(s) changed?

_____ 7. If combination is recorded, where is it stored?

_____ 8. Total number of employees possessing combination?

_____ 9. Review procedures for securing sensitive items, i.e. monies, precious metals, high dollar value items, narcotics, etc.?

_____10. Who performs locksmithing function for the facility?

_____11. Is a key inventory periodically taken?

_____12. Are locks changed when keys are lost?

Comments:

VI. ALARMS

_____ 1. Does this facility utilize any alarm devices?
Total number of alarms?

Type	Location	Manufacturer	Remarks

_____ 2. Are alarms of central station type connected to police department or outside guard service?

_____ 3. Is authorization list of personnel authorized to "open & close" alarmed premises up to date?

_____ 4. Are local alarms used on exit doors?

_____ 5. Review procedure established on receipt of alarm?

_____ 6. Is closed circuit TV utilized?

   Comments:

VII. GUARDS/SECURITY CONTROLS

_____ 1. Is a guard service employed to protect this facility?

   If yes. Name:_____ No. of guards_____ No. of posts_____

_____ 2. Are after hours security checks conducted to assure proper storage of classified reports, key controls, monies, checks, etc.?

_____ 3. Is a property pass system utilized?

_____ 4. Are items of company property clearly identified with a distinguishing mark that cannot be removed?

_____ 5. Review guard patrols & frequency?

_____ 6. Are yard areas and perimeter areas included in guard coverage?

_____ 7. Are all guard tours recorded?

_____ 8. Are package controls exercised re packages brought on or off premises?

_____ 9. Does facility have written instructions for guards?

_____10. What type of training do guards receive?

_____11. Are personnel last leaving building charged with checking doors, windows, cabinets, etc.?
   Record of identity?

_____12. Are adequate security procedures followed during lunch hours?

   Comments:

VIII. EMPLOYEE AND VISITOR CONTROLS

_____ 1. Is a daily visitors register maintained?

_____ 2. Is there a control to prevent visitors from wandering in the plant?

_____ 3. Do employees use identification badge?

_____ 4. Are visitors issued identification passes?

_____ 5. What type of visitors are on premises during down hours and weekends?

_____ 6. Does any company's employees other than _____ have access to facility?
   List Company Names   Type Service Performed
   _____   _____

_____ 7. Are controls over temporary help adequate?

Comments:

IX. PRODUCT CONTROLS (Shipping and Receiving)

_____ 1. Are all thefts or shortages or other possible problems, i.e., anonymous letters, crank calls, etc. reported immediately?

_____ 2. Inspect and review controls for shipping area.

_____ 3. Inspect and review controls for receiving area.

_____ 4. Supervision in attendance at all times?

_____ 5. Are truck drivers allowed to wander about the area?
   Is there a waiting area segregated from product area?
   Are there toilet facilities nearby?
   Water cooler?
   Pay telephone?

_____ 6. Are shipping or receiving doors used by employees to enter or leave facility?

_____ 7. What protection is afforded loaded trucks awaiting shipment?

_____ 8. Are all trailers secured by seals?

_____ 9. Are seal numbers checked for correctness against shipping papers? "In" and "Out"

_____10. Are kingpin locks utilized on trailers?

_____11. Is a separate storage location utilized for overages, shortages, damages?

_____12. Is parking (employees and visitor vehicles) prohibited from areas adjacent to loading docks or emergency exit doors?

_____13. Is any material stored in exterior of building?
   If so how protected?

_____14. Are trailers or shipments received after closing hours?
   If so how protected?

_____15. Are all loaded trucks or trailers parked within fenced area?

_____16. Review facility's product inventory control.

	Loss	Breakage	Returns
Average Monthly			

_____17. Review controls over breakage.

Comments:

X. MONEY CONTROLS

_____ 1. How much cash is maintained on the premises?

_____ 2. What is the location and type of repository?

_____ 3. Review cashier function.

_____ 4. What protective measures are taken for money deliveries to facility? To bank?

_____ 5. If armored car service utilized, list name and address.

_____

_____ 6. Does facility have procedure to control cashing of personal checks?

_____ 7. Are checks immediately stamped with restricted endorsement?

_____ 8. Are employee payroll checks properly accounted for and stored in a locked container (including lunch hours) until distributed to the employee or his supervisor?

Comments:

XI. PROPRIETARY INFORMATION

_____ 1. What type of proprietary information is possessed at this facility?

_____ 2. How is it protected?

_____ 3. Is "_____ Restricted" marking used?

_____ 4. Are safeguards followed for paper waste, its collection and destruction?

_____ 5. Are desk and cabinet tops cleared at end of day?

_____ 6. Is management aware of need for protecting proprietary information?

Comments:

XII. OTHER VULNERABILITIES

_____ 1. Trash pick ups. (Hours of pick ups, control of contractor, physical controls).

_____ 2. Scrap operations. (Physical controls of material and area, control over scrap pick ups, etc.).

_____ 3. Other.

Comments:

XIII. PERSONNEL SECURITY

_____ 1. Are background investigations conducted on employees handling products?
       Handling cash?
       Engaged in other sensitive duties?
       Supervisory position?
       All employees?

_____ 2. If so, who conducts background investigation?

_____ 3. Are new employees given any security or other type of orientation?

_____ 4. Do newly hired employees execute a corporate briefing form for inclusion in their personnel file?

_____ 5. Are exit interviews conducted of terminating employees?

_____ 6. Is a program followed to insure return of keys, credit cards, I.D. cards, manuals, and other company property?

GENERAL COMMENTS

Appendix C

SECURITY DEPARTMENT
LOSS PREVENTION CHECK LIST

STORE NO._____

Legend:
A—Satisfactory
B—Corrected During
Inspection
C—Requires Corrective
Action

INSTRUCTIONS
All areas applicable to this store are to be inspected during course of month. Indicate condition by marking "X" in appropriate box. Areas requiring corrective action should be reviewed with store management prior to turning in check list. Use Part II for explanatory details.

A	B	C	PART I	Date Inspected
			Securement of Theft Prone Merchandise 1. Calculators properly secured to display fixture, (Dept. 13)	
			2. Display typewriters are secured (Dept. 15)	
			3. Lockbars in Suit Dept. are in place & kept locked.	
			4. Men's Leathers are chained or are under lockbar.	
			5. Stoplifters are used on higher priced RTW fixtures adjacent to street exits.	
			6. Stoplifters are utilized on floor fixtures in Dept. 9-48-61.	
			7. Men's Accessories showcase are kept locked (Dept. 105)	
			8. Ciani counter fixtures are secured to counter (Dept. 20)	
			9. All portable TV's are secured with cord locks. (Dept. 72)	

A	B	C		Date Inspected
			Physical Security 10. Firelocks on stairwells & emergency exits have been tested.	
			11. Locks & electric latch on vault man-trap are functional.	
			a. Access list to Cashier area and vault are current & posted.	
			12. All Ultrasonic Alarms (Fine Jewelry, Furs, Escalators) walk tested.	
			13. Restricted area signs posted at inner hallways to dock & back areas.	
			14. Rear exit door from restaurant kept locked at all times & key under control of restaurant manager.	
			15. Alteration Room secured whenever unoccupied or closed.	
			16. Watchman is assigned to employee door after employee system is turned off.	
			Merchandise Counts 17. TV counts taken prior to store opening.	
			18. Daily suit counts accurately taken on timely basis.	
			19. Fur counts are taken at store opening & closing & include all furs in stockroom (include holds & customer repairs).	
			Stockrooms 20. All stockroom doors located in fitting room banks are equipped with self locking hardware.	
			21. Lockable stockrooms are kept locked.	
			22. Stockrooms utilize only one entrance (alternate entrances are closed off).	
			23. TV stockroom is completely enclosed, adequate in size & kept locked at all times.	
			24. Keys to TV stockroom controlled only by Dept. Manager & store management.	
			25. Stockroom Doors are not propped or tied open unless employee continuously present.	
			Dock, Wheeler, Basement, Penthouse 26. Access to customers is secured during night hours & Sundays.	
			27. High theft type merchandise not stocked in open Wheeler areas.	
			28. Transfer trailer seals are stored in secure location.	
			29. Customer pickup log is properly maintained (include vehicle license number).	
			30. Dock doors kept closed when loading operation not in progress.	
			31. Employee always present when dock doors open.	
			32. Marking room kept locked when unattended.	
			33. Trailers are secured during lunch breaks & end of day.	
			Fitting Rooms 34. Each fitting room & fitting room hallway is properly signed with signs in good condition (1-2-3, policy statement & shoplift picture).	

A	B	C		Date Inspected
			35. Garment limit sign clearly visible in each entrance foyer area.	
			36. Fitting room doors properly cut (top & bottom) to permit surveillance, or vented in proper direction.	
			37. Fitting room has only one entrance/exit.	
			38. Store is equipped with adequate supply of fitting room discs and disc holders.	
			39. Fitting rooms are not used as stock storage areas unless modified to provide secure storage.	
			40. Fitting room wall abutting stockroom has floor to ceiling wall.	
			41. Fitting room walls of one department abutting departments fitting room bank to have floor to ceiling wall.	
			42. Merchandise is cleared out of all fitting rooms on timely basis.	
			High Shrinkage Departments 43. Costume Jewelry	
			a. Counter fixtures with higher priced merchandise located closer to register stand & in unobstructed view.	
			b. All display cases & drawers are equipped with locks & all locks are functional.	
			c. All display cases & drawers are locked each night (verify).	
			d. Countertop fixtures are covered with dust covers at store closing.	
			44. Wallets (Dept 86) displayed only one deep & in manufacturer's box when reserve stock can be kept in dept. area.	
			45. Furs	
			a. No fur piece in excess of $995 is displayed on selling floor, except on mannequin with arms.	
			b. Furs not displayed on more than 4 floor fixtures & each display is balanced.	
			c. Keys to fur vault during day are retained in register or on salesclerks' person. (Not in unlocked desk)	
			d. Furs are removed from floor display and secured when employee with 60 number is away from Dept. (including breaks, lunch, etc.)	
			e. Fur dept. & PBX employees are familiar with procedure to be followed re: Fur Dept. buzzer & buzzer is functional.	
			46. Cosmetic testers are properly marked.	
			47. Dept 106 Jewelry is displayed in area clearly visible from register stand.	
			48. Dept 102 Jewelry is displayed in area clearly visible from register stand.	
			49. All TV's not displayed on selling floor are stored in TV stockroom and placed in stockroom immediately upon receipt.	

A	B	C		Date Inspected
			Administration	
			50. Security Officer is regular attendee at store's Group Manager meetings.	
			51. Store management is kept informed of attempted theft activity (burns).	
			52. Store management/Security periodically monitors employee door during early shift breaks, i.e. House-keeping crews, Stock, Display.	
			53. Store management representative periodically reviews loan books to insure accuracy.	
			a. Audit loan book for 2 depts. and note findings in Part II.	
			54. Over & short records are kept posted on a current basis.	
			a. How far behind are postings?	
			b. Name of store representative specifically designated to be responsible to monitor Over & Shorts:	
			55. Periodic store walk-throughs are conducted by either Senior or Security Officer with a member of store management.	
			a. Date of last joint walk-through	
			b. Note in Part II conditions noted in walk-through not yet corrected.	
			56. Are checker schedules filled as planned?	
			a. Store includes Checkers in break relief schedule.	
			57. Are Checkers hired on non-sell number where possible instead of using contingent sales personnel?	
			58. Are proper employee lockering techniques followed?	
			a. Spot check 3 employee sealed lockered packages.	
			59. Are employees wearing their store badges?	
			a. Take count of all employees visible on one tour of the store. Employee's badged_____Not badged_____.	
			60. Is store initiating loss reports for suspected losses?	
			a. Turn in quantity of Kimball Tickets recovered from fitting rooms.	
			Qty. turned in_____Loss Reports prepared_____ (explain in Part II)	

PART II

Use this section to elaborate on any answer in Part I or to make additional comments in other areas. Continue comments on reverse side. Attach additional sheets if needed.

PART III

List by item number the inspection items reviewed with store management.

Member of Store Management with Whom Reviewed: _____

_____   _____   _____
Name of Security Employee   Signature   Date of Report
Conducting Inspection

INDEX

Index

Other Security World Books of Interest . . .

LOSS PREVENTION: CONTROLS AND CONCEPTS
By Saul D. Astor (273 pp.)
Thought-provoking examination of loss prevention by a leading security professional, zeroing in on controls that work, and those that don't. Includes "Astor's Laws" of loss prevention; early warning signals of employee dishonesty; special retail problems.

MANAGING EMPLOYEE HONESTY
By Charles R. Carson (230 pp.)
Systematic approach to total accountability in any business. Draws on case histories and sound business management principles to show how to hire honest employees; control honesty in all phases of business.

INTRODUCTION TO SECURITY
(Revised Edition)
By Gion Green and Raymond C. Farber (336 pp.)
Comprehensive introduction to the history, nature and scope of security in modern society. Basic principles of physical security, risk analysis, internal loss prevention, fire prevention and safety.

COMPUTER SECURITY
By John M. Carroll (400 pp.)
Total program of protection for EDP systems and facilities, from basic physical security to sophisticated protection of hardware, software and communications.

SUCCESSFUL RETAIL SECURITY
An Anthology (303 pp.)
Over 25 top security and insurance professionals pool expertise in retail loss prevention, including employee theft, shoplifting, robbery, burglary, shortages, fire protection, insurance recovery.

ALARM SYSTEMS & THEFT PREVENTION
By Thad L. Weber (385 pp.)
Definitive treatment of alarm systems and problems, in laymen's terms. Strengths and weaknesses of alarm systems, how they are attacked by criminals, and techniques required to defeat these attacks.

INTERNAL THEFT: INVESTIGATION & CONTROL
An Anthology (276 pp.)
Top security professionals analyze employee dishonesty and how to control it. 25 chapters on: Why Employees Steal; Executive Dishonesty; Embezzlement; Undercover Investigation; Pre-Employment Screening; more.

OFFICE & OFFICE BUILDING SECURITY
By Ed San Luis (295 pp.)
Security solutions for offices, high-rise buildings and personnel. Analyzes external crimes and systems of defense, internal crime and protection against specific dangers.

CONFIDENTIAL INFORMATION SOURCES:
PUBLIC & PRIVATE
By John M. Carroll (352 pp.)
Unique guide to public and private personal records. Reveals what information is on file, how it is gathered, who has access, how to identify the unknown person.

BOMB SECURITY GUIDE
By Graham Knowles (157 pp.)
Step-by-step emergency program against bomb threats and letter bombs. Covers device recognition, telephoned bomb threat procedures, emergency response including evacuation, search and safety rules.

AIRPORT, AIRCRAFT & AIRLINE SECURITY
By Kenneth C. Moore (356 pp.)
Definitive study of every aspect of air traffic security, from hijacking to predeparture screening and baggage handling; airport physical protection; credit card fraud, internal theft and investigation.

HOSPITAL SECURITY
By Russell L. Colling (384 pp.)
Complete protection of people and property in health care facilities. Practical, detailed programs to deal with hospital vulnerabilities including theft of drugs, assault, fire, disaster, internal theft.

HOTEL & MOTEL SECURITY MANAGEMENT
By Walter J. Buzby II and David Paine (256 pp.)
Loss prevention in the hotel industry and protective measures for the hotel or motel, large or small. Includes theft, holdup, fraud, fire, restaurant and bar security, innkeeper's liability for injury to guests.

In addition to its hard cover books on security subjects, Security World Publishing Company publishes *Security World* and *Security Distributing & Marketing (SDM)* magazines; produces booklets, manuals and audio tape cassettes on security; and sponsors the International Security Conference. Books and other materials are available from Security World Publishing Co., Inc., 2639 South La Cienega Blvd., Los Angeles, California 90034.